ISBN 978-1-330-48296-4
PIBN 10067759

English
Français
Deutsche
Italiano
Español
Português

www.forgottenbooks.com

Mythology Photography **Fiction**
Fishing Christianity **Art** Cooking
Essays Buddhism Freemasonry
Medicine **Biology** Music **Ancient
Egypt** Evolution Carpentry Physics
Dance Geology **Mathematics** Fitness
Shakespeare **Folklore** Yoga Marketing
Confidence Immortality Biographies
Poetry **Psychology** Witchcraft
Electronics Chemistry History **Law**
Accounting **Philosophy** Anthropology
Alchemy Drama Quantum Mechanics
Atheism Sexual Health **Ancient History**
Entrepreneurship Languages Sport
Paleontology Needlework Islam
Metaphysics Investment Archaeology
Parenting Statistics Criminology
Motivational

MODERN SHIPBUILDING

AND

THE MEN ENGAGED IN IT:

A REVIEW OF RECENT PROGRESS IN STEAMSHIP DESIGN AND CON-
STRUCTION, TOGETHER WITH DESCRIPTIONS OF NOTABLE

SHIPYARDS, AND STATISTICS OF WORK DONE IN

THE PRINCIPAL BUILDING DISTRICTS.

BY

DAVID POLLOCK,

NAVAL ARCHITECT.

With Portrait and Biographical Notes of Eminent Shipowners,
Shipbuilders, Engineers, and Naval Architects;
also, Views of Notable Ships.

LONDON: E. & F. N. SPON, 125 STRAND.
NEW YORK: 35 MURRAY STREET.

1884.

D

PREFACE.

THE great activity in shipbuilding and marine engineering during recent years, and the substantial progress, both in science and practice, which has marked the period, have often formed the subject of articles in the technical and daily press, and of papers read before professional institutions. So far as I am aware, however, no single work dealing historically with modern shipbuilding in a way at once trustworthy and popular, and in a form handy and accessible, has yet been published. The present work aims at supplying this want. In undertaking it originally, I felt encouraged by the acceptance which various articles, contributed to the columns of the *Glasgow Herald, The Engineer, The Steamship, Iron,* &c., had met with from many whose good opinion I had reason to value highly. With the kind permission of the proprietors of the above journals, I have made use to some extent of the articles in question—but largely amplified and corrected—in preparing the following pages.

The work is concerned exclusively with shipbuilding for the merchant marine, and no attempt is made to trace the progress connected with naval shipbuilding, although some of the many important influences which the one exerts upon the other have been indicated. Even as thus defined and restricted, the field of review is so vast that the limits which I had determined should bound the work with respect to price, and consequently with respect to size, have compelled me to treat briefly and in a general way many matters which it might have been of interest to enlarge upon. The list of authoritative papers and lectures to which readers can at first hand refer—given at the end of each chapter—may, it is hoped, compensate to some extent for these deficiencies.

The book being mainly historical, originality in the strict sense of the term cannot, of course, be urged for much of the contained matter; but efforts have been made throughout to present trustworthy statements of the very latest steps in advance. This is specially true of the chapter on scientific progress. My object, however, having been more to enlighten general readers than to seek to interest or inform professional ones, it is perhaps wanting in the scientific fulness needed to give it special value, viewed from the standpoint of the trained naval architect.

While the biographies and portraits given throughout the book may be considered fairly representative of those who as shipbuilders, shipowners, naval architects, or marine engineers have made their influence felt on the world's mercantile marine during the period of review, the collection by no means includes all who are deserving of such notice. The subjects of portraiture are all in life, and actively engaged in their respective spheres of labour. The diffidence generally evinced by them in consenting that their likenesses and the note of their professional career should be given, has made my task one of difficulty. What may be called the over-diffidence of a few, originally selected for portraiture, has to some extent occasioned the incompleteness now commented upon.

As further accounting for the limitations of the present work, I think it fitting to add that the preparation of the whole book, including the task of seeing it through the press, has devolved upon me at a time when the ordinary intervals of respite from daily business have had to suffice for its accomplishment.

My best thanks are due to those firms and individuals to whom I had to appeal for statistics and other particulars, for their generally ready and courteous attention to my requests.

DAVID POLLOCK.

DUMBARTON, *November*, 1884.

CONTENTS.

CHAPTER I.

RECENT PROGRESS IN STEAMSHIP CONSTRUCTION. PAGE ·

CHAPTER II.

SPEED AND POWER OF MODERN STEAMSHIPS.

CHAPTER III.

SAFETY AND COMFORT OF MODERN STEAMSHIPS.

CHAPTER IV.

PROGRESS IN THE SCIENCE OF SHIPBUILDING.

CHAPTER V.

PROGRESS IN METHODS OF SHIPYARD WORK.

CHAPTER VI.

DESCRIPTIONS OF SOME NOTABLE SHIPYARDS.

CHAPTER VII.

OUTPUT OF TONNAGE IN THE PRINCIPAL DISTRICTS.

CHAPTER VIII.

THE PRODUCTION OF LARGE STEAMSHIPS.

APPENDIX.

CALCULATING INSTRUMENTS.

PORTRAITS AND BIOGRAPHICAL NOTES.

VIEWS OF NOTABLE STEAMSHIPS.

ERRATA.

PAGE 11.—Thirteenth line from top: for 1883 read 1884.
PAGE 81.— Fourth line from top : for "a single trial" read "one or two trials."
PAGE 163.—Fourth line from top: for 1884 read 1845.
PAGE 187.—Third line from top: for "fluctuations" read "fluctuation."
PAGE 200.—Dimensions of "City of Rome": for 546 by 52 by 58¾ read 546 by 52 by 38¾.

"INTO a ship of the line man has put as much of his human patience, common sense, forethought, experimental philosophy, self-control habits of order and obedience, thoroughly wrought handwork, defiance of brute elements, careless courage, careful patriotism, and calm expectation of the judgment of God, as can well be put into a space 300 feet long; by 80 feet broad."—*Ruskin.*

"If any body of men have just cause to feel pride in their calling, and in the fruits of their labour, shipbuilders have. If we look at the magnitude of the operations of building, launching, engining, and completing a modern passenger ship of the first rank, and regard the multiplicity of the arrangements and beauty of finish now expected, and then think this structure has to brave the elements, make regular passages, convey thousands of human souls, and tens of thousands of tons of merchandise every year across the ocean, in storm or calm, we cannot but feel that they are occupied in useful human labour. But more than this, there is a public sentiment surrounding ships that no other mechanical structures can command Beautiful churches, grand buildings, huge structures of all kinds have a certain interest pertaining to them, but it is different in kind from that which surrounds a ship. The former are fixed, immovable, inert; the ship is here to-day and gone to-morrow, building up a history from day to day with a reputation as sensitive as a woman's to calumny, and like her consequently often a bone of contention as well as an object of admiration."—*William John.*

MODERN SHIPBUILDING.

CHAPTER I.

RECENT PROGRESS IN STEAMSHIP CONSTRUCTION.

THE achievements in shipbuilding and marine engineering within recent years may be said to borrow lustre from one particular feat of past times. The *Great Eastern* undoubtedly furnished, in large measure, the experience that has recently been causing so great a change in the tonnage of our mercantile marine. Commercially, as is well known, that huge vessel—"Brunels' grand audacity," she has been called—has all along proved a lamentable failure. It has been stated on good authority that between 1853—the year in which the contract for her was entered into—and the year 1869, no less than one million sterling had been lost upon her by the various proprietors attempting to work her. Financially, indeed, she may be said to have proved the "Devastation" of the mercantile marine. Although at various times in her long life-time she has unquestionably done most useful service in sub-marine cable-laying—service, indeed, which, but for her, could not well have been accomplished—these times of usefulness have been far outbalanced by her long periods of inactivity.

Apart from commercial considerations, however, this premier leviathan still stands out as a wonder and pattern of naval construction. In her admirably-conceived and splendidly-wrought stuctural arrangements—due to the joint labours of the late Mr I. K. Brunel and Mr J. Scott Russell—she possesses as successful an embodiment of the dual quality of "strength-with-lightness" as can be found in any subsequent ocean-going merchant ship. She was, if not the first, certainly the greatest embodiment of the longitudinal system of construction, and in virtue of this, as well as of her phenomenal proportions, she represents, alone, more of the intrepidity and skill essential to thorough progress, than are exhibited by combined hosts of the " departures " of recent times.

Despite the far-reaching views of the eminent designer, those changes which have since taken place in the essential conditions for successful ocean navigation eluded his vision. Owing to the opening of coal mines in almost all parts of the world, it is now no longer necessary nor desirable that a steamer should be capable of carrying coals for a return voyage, either from India or Australia—this being the dominant and regulating condition in the *Great Eastern's* design.   Further, the improvements in marine engineering, represented by the greater possible economies in coal consumption and the fuller utilization of steam, which have since been effected, have rendered the great ship inefficient and obsolete.   In short, Brunel and his financial supporters were ahead of their time, and failed to appreciate the law of progress, now better understood— " invention must wait on experience."

The urgent demands of our broader civilisation, improvements in navigation, the spread of population in new colonies and over wider continents, and, above all, the fresh accessions of experience and invention, are forces which now impel shipowners to increase the dimensions of their vessels, and shipbuilders to carry out the work.   Each year the contrasts as to dimensions between the first leviathan and her later sister grow less and less.   The completion within the past few years of such monster merchant ships as the *Servia,* the *City of Rome,* the *Alaska,* and the *Oregon,* and the forward state of the *Etruria* and *Umbria,* two remarkable steamships, building on the Clyde for the Cunard Company, constitute an epoch in the history of our mercantile marine, and give colourable justification to the belief sometimes expressed, that the proportions of the *Great Eastern* will in time be surpassed.

The feasibility—in a scientific sense—of ships growing in proportions commensurate with the growth of commerce and traffic, has often been commented upon.   The whole tendency of our time is towards the aggregation of effort : the massing of

PORTRAIT
AND
BIOGRAPHICAL NOTE.
———
JOHN BURNS.

JOHN BURNS, F.R.A.S., F.R.G.S.

CHAIRMAN OF THE CUNARD STEAMSHIP COMPANY.

BORN at Glasgow and educated at the University in that city. At an early age became a partner in the firm of G. & J. Burns, which was founded in 1824 by George Burns (his father) and James (his uncle), also in the Cunard Steamship Company, of which gigantic concern, as is well known, his father, with Samuel Cunard and David M'Iver, were the founders in 1839. From the first Mr BURNS earnestly addressed himself to the responsibilities of his important position, and finding able coadjutors in his other partners in the Cunard Company, has carried on the concerns of that great Steamship Line so as to enhance its reputation and maintain first place in the Atlantic Mail Service. In 1880, forty years after its formation, the Company was transformed into a public corporation, with Mr BURNS as chairman. The fleet now consists of 37 steamers, representing over 110,000 tons, or a money equivalent of nearly £3,000,000, and giving employment to an enormous number of persons. While everything is done on board to ensure speed and comfort, the main consideration, to which all others are made subservient, is *safety*. First-class vessels, unstinted equipment, carefully-selected officers and men, combined with close personal supervision, are the means used to attain this end, and that it is attained marvellously is matter of world-wide fame. Apart from his able management of the Cunard fleet, Mr BURNS has not allowed the affairs of his Home Services between this country and Ireland and elsewhere, to suffer in any particular, but in his hands these concerns have flourished and the trade greatly increased. The services are conducted by a splendid fleet of mail steamers, now belonging exclusively to Mr BURNS, quite irrespective of the Cunard fleet, and which, for speed, safety, and unfailing regularity of departure and arrival, are probably unsurpassed. As representing the Cunard Company, and also as a private ship-owner, Mr BURNS has taken frequent and conspicuous part in the discussion of those great matters which concern the maritime interests of this country. Has often been called upon to give evidence before Select Committees of the House of Commons on shipping affairs. Was amongst the first to recommend to Government the desirability of fitting merchant steamships so as to be available in times of war. Is Deputy-Lieutenant of Lanarkshire, and Magistrate for the counties of Lanark and Renfrew. Evinces unbounded interest in the commercial and social well-being of his native city, numerous benevolent institutions in great measure owing their existence to his hearty munificence. His residence of Castle Wemyss, on the Clyde, is frequently the abode of the famous of this and other countries.

John Burns

capital and labour. A vessel of five thousand tons can be built cheaper than five vessels of one thousand tons. In the manning and working of ships there is a still more striking economy, *e.g.*, one captain instead of five, and so on throughout the staff of officers, engineers, stewards, and crew. Not only so, but long ships can be propelled at greater speeds than short ones, the whole conditions of construction, engines, and propellers being considered. Mr Robert Duncan, in his presidential address before the Society of Engineers and Shipbuilders in Glasgow in 1872, declared:—"Looking forward one generation, and measuring the future by the past, I think it is not problematical that we shall see steamers of eight hundred feet long the ferryboats of two oceans, with America for their central station, and Europe and Asia for their working termini." Even since that was uttered, eleven years ago, we have approached, in solid practice, the limit thus laid down, by 150 feet at least. Three years previous to Mr. Duncan's address, vessels exceeding four hundred feet were not afloat, with the notable exception already referred to; now, there are few merchant fleets of any pretensions engaged in ocean traffic which do not include vessels over or approaching four hundred feet, and it is even no great boast that vessels close on six hundred feet are afloat and in active service.

As better illustrating the growth in dimensions of merchant steamships, the Figs. on the following page may prove interesting. They show, all to the same scale, a number of representative steam vessels from the *Comet* downwards.

Along with the change or evolution in the sizes and types of merchant vessels, important modifications in their structural arrangement have of late years been effected, and it is to the constant progress being made in these matters—to the skill and intrepidity which are brought to bear on their execution, and to the readiness with which our shipowners recognise their importance and value—that the maintenance of our mercantile supremacy is largely owing. An American

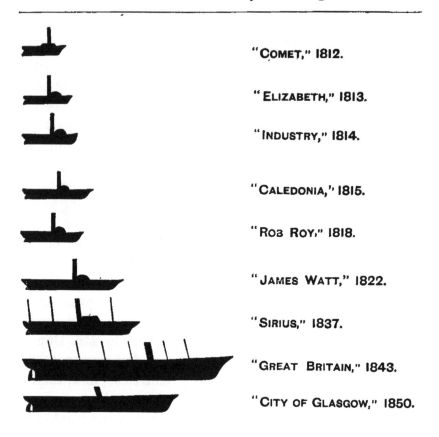

"COMET," 1812.

"ELIZABETH," 1813.

"INDUSTRY," 1814.

"CALEDONIA," 1815.

"ROB ROY," 1818.

"JAMES WATT," 1822.

"SIRIUS," 1837.

"GREAT BRITAIN," 1843.

"CITY OF GLASGOW," 1850.

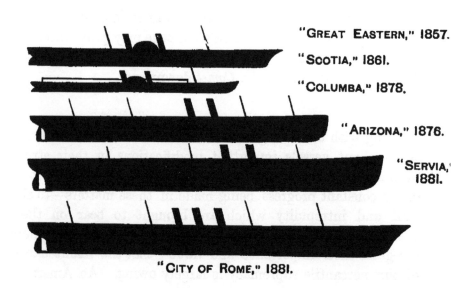

"GREAT EASTERN," 1857.

"SCOTIA," 1861.

"COLUMBA," 1878.

"ARIZONA," 1876.

"SERVIA," 1881.

"CITY OF ROME," 1881.

journal, writing a few years ago on this subject—perhaps with more of taunt for the conceit and self-sufficiency evinced by its own country than of adulation for the ability and enterprise displayed by ours—said:—

" In the whole world there is no place whatever that can in any degree compare with the Clyde for either extent or quality of steamship building ; and at this moment an indisputable verification can be adduced, for between American and European ports there are at the present time something like a score of steam navigation companies, doing an immense passenger and carrying trade, with vessels of great power and magnificence, and notwithstanding the variety of trade nationalities, at least two-thirds of the vessels employed were built and equipped on the Clyde ; and more—unless there has very recently been a change, there is not an American steam company in the whole Atlantic trade. With a run of about fifty years to try it, and after many unsuccessful attempts, the Americans have utterly failed to sustain permanent competition. All the British companies have prospered beyond any probable anticipation clothed with reason. The Cunard Company, starting with four vessels some forty years ago, have now twenty times that number. What is this something which enables Europeans to so far outstrip the Americans in a competitive traffic so as to exclude them from the merest show in the largest steam trade in the world ? A baneful, overweening, and ignorantly selfish conceit invariably leads to disastrous results, and a nation given over to the fulmination of concentrated boast cannot fail to be suffocated with foolery of its own making."

This is doubtless the outcome of a vicious antipathy—natural in the circumstances—to those stringent and over-reaching laws which forbid that ships built away from America shall sail under the American flag, or enjoy the pertaining privileges. American shipbuilders thus secured from the encroaches of foreign competition, have enjoyed their own pace, but at too great a sacrifice. Preferring to take the material most at hand, the manipulation of which they well understood, they have allowed their wood age to be dove-tailed thirty years into our iron one, with the other result that America now occupies as unimportant a place in the traffic of the sea, as the above quotation indicates.

Evidences are not wanting, however, to show that America is at least endeavouring, in some respects, to be abreast of the times, and that she has brought herself to acknowledge and follow the lead of this country. In this connection, the four new vessels presently being constructed for the U.S. Navy may be shortly referred to. The vessels comprise three cruisers and one despatch boat, all of which are being built by Mr John Roach, of Chester, Pa., the material employed in their construction being mild steel of American manufacture Twin screws will be employed for the propulsion of the largest vessel—the *Chicago*—which is to be 315 feet long between perpendiculars, 48 feet beam, and 34 feet 9 inches moulded depth to spar deck. The other vessels are the *Boston* and the *Atalanta*, single screw cruisers of 270 feet length; and the *Dolphin*, single screw despatch boat, of 250 feet length and high speed.

In almost every feature except machinery these new American naval vessels strongly resemble Government vessels of recent British build, a circumstance for which there is little difficulty in accounting, as it is well known the naval authorities in the States have within recent times been recruited by young American naval architects educated in our Naval College at Greenwich, and consequently steeped in British naval practice. This and other facts, such as the visit of a technical commissioner of the States' navy, two years ago, to our naval and mercantile shipyards—upon which he has since fully reported—leave one in no doubt as to the source of coincidence in design and structure.

The subject of America's position as a shipbuilding and shipowning country has involved reference to wood shipbuilding, but to revert at any length to this topic in a work dealing with modern progress in British shipbuilding, the bulk of which is written of and for industrial and commercial centres where wood shipbuilding has been long entirely tabooed, is quite unnecessary. Doubtless, however, the amount of wood and composite building still carried on in the minor seaports of the United Kingdom, and in several of the British possessions, is of sufficient importance to demand some reference. As the present position of affairs in this con-

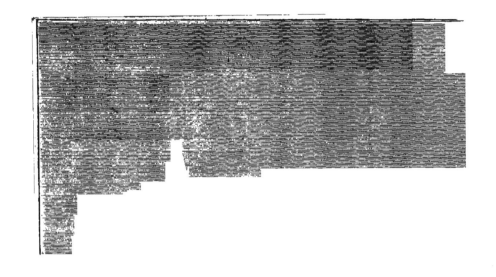

S.S. UMBRIA.—CUNARD LINE.

| LENGTH, | ... | ... | 500 ft. 0 in. | DEPTH, | ... | ... | 40 ft. 0 in. |
| BREADTH, | ... | ... | 57 ft. 0 in. | TONNAGE (GROSS),... | | | 7,718 tons. |

BUILT BY MESSRS. ELDER & CO., 1884.

nection is briefly and forcibly illustrated by statistics compiled and issued by the British Iron Trade Association, two tables taken from this source may be given, the subject thereafter being finally departed from :—

Tonnage of Vessels constructed and registered in the United Kingdom of Iron, Steel, and Wood respectively, in each of the years 1879 to 1883, with Percentage of Total Tonnage constructed in Iron and Steel.

Year.	Gross Tonnage of Vessels built of		
	Iron and Steel.	Wood.	Excess Tonnage in Iron and Steel.
1879	484,636	26,186	458,450
1880	525,568	19,938	505,630
1881	730,686	18,107	712,579
1882	913,519	14,850	898,669
1883	1,012,735	15,202	997,533
Totals,...	3,667,144	94,283	3,572,861

Tonnage of Wooden Vessels registered in the United Kingdom which were Lost, Broken up, &c., during each of the years 1879 to 1883, with Tonnage of Wooden Vessels built and registered in the United Kingdom during the same period.

Year.	Tonnage of Wooden Vessels.		Excess of Vessels lost over those built.
	Lost.	Built.	
1879	149,828	26,186	123,642
1880	173,065	19,938	153,127
1881	170,283	18,107	152,176
1882	166,809	14,850	151,959
1883	144,138	15,202	128,936
Totals,...	804,123	94,283	709,840

Whence it appears that while 709,840 tons of the 1,779,112 tons of ships removed from the register during the last five years were wooden vessels, only 94,283 tons of the 3,667,144 tons built and registered in the United Kingdom during the same period were constructed of that material. In other words, wooden ships represent 45 per cent. of the total losses, while they only represent 2·5 per cent of the total tonnage built and added to the register during the five years in question.

Just as the introduction or general adoption of the compound engine marked an epoch in the history of shipbuilding and marine propulsion, so now the introduction of "mild

steel" or "ingot iron" as a material for shipbuilding, together
with the more extended adoption of water ballast, and the
rapid development of the continuous cellular system of construction, may be said to constitute a fresh starting-point in
the history of the industry.

Although the introduction of steel as a material for shipbuilding dates at least as far back as 1860, its use has been
but partial or occasional until within very recent times. The
uncertainty as to quality, the frequent great disparity between
pieces cut from the same plate, and the special care needed in
the manipulation, prevented its general adoption. With the
highly improved "mild steel," however, first manufactured in
France, and applied to shipbuilding purposes there about nine
years ago, and subsequently introduced into this country,
began the more extended adoption of steel, which every day,
or with every accession to experience, is displacing iron.

The facts relating to the introduction into this country of
mild steel for shipbuilding purposes, may be briefly recounted.
In the latter end of 1874, Admiral Sir W. Houston Stewart,
Controller of the British Navy, and Mr. N. Barnaby, Director
of Naval Construction, availed themselves of the opportunity
to observe and study the use of steel in the French dockyards
of Lorient and Brest, where three first-class armour-plated
vessels were then being built of steel throughout, supplied
from the works at Creusot and Terrenoire. Mr. Barnaby, at
the meetings of the Institution of Naval Architects in March
following, gave an account of his observations during this visit,
and pointed out clearly and precisely to the steel-makers of
Great Britain all the indispensable conditions which would
have to be met and satisfied by steel for shipbuilding, so that
it could be used with confidence in the construction of the
largest vessels. Before the end of 1875, the Landore-Siemens
Company was enabled to fulfil these conditions, and the
Admiralty contracted with them to supply the plates and
angles necessary for the construction of two cruisers of high
speed—the *Iris* and the *Mercury*. The material involved in
this contract was steel obtained by the Siemens-Martin process.

Shortly after this the Bolton Steel Company was in its turn able to produce by the Bessemer process plates and angles, satisfying all the requisite conditions. The Steel Company of Scotland, Butterly Company, and other important works, also entered into the same business, and operations are still going on in various parts of the country connected with the formation of new works, and the perfecting of other processes.

The steel furnished by these different works, subjected as it has been to systematic and severe tests continually applied, is now possessed of the qualities of ductility, malleability, and homogeneity, which render its employment in shipbuilding not only permissable but highly desirable. Its good and reliable qualities have been admitted by the Constructors of the Navy, the Officers of the Board of Trade, of Lloyd's, and of the Liverpool Registries, as well as by all the most competent authorities. The experience of all who have practical dealings with the material in the shipyard is that it entirely satisfies—even more than iron—all the requirements of easy manipulation. The confidence with which it can be relied on, as to its certain and uniform qualities, places it on a much higher level than the steel formerly manufactured; and its superiority over the best wrought iron as regards strength and ducility renders it a highly preferable material.

While doubt exists, however, as to the adoption of steel for shipbuilding being commercially advantageous; there must be hesitancy on the part of shipowners and others concerned. Although, since its introduction, mild steel has been greatly reduced in price, the first cost of a steel ship is still somewhat over that of an iron one, even after the reduction in weight of material is made, which the superiority of steel permits of. It has been shown that, about two years ago, a spar-decked steamer, of 4,000 tons gross, built in steel, as against a similar vessel built in iron, entailed an excess in cost of £3,570. The advantages, however, which accrue from the change, both immediate and in the long-run, make the gain clear and considerable. Steel ships have been built with scantlings reduced one-fourth or one-third, and in some early cases even *one-half,* from what would have been considered requisite had iron been employed. Some authorities, not unnaturally, questioned the wisdom of accrediting steel with all the qualities which make such sweeping reductions justifiable. Except in vessels for river or passenger service, however, this is much in advance of the reductions obtained in ordinary modern practice.

The reductions allowed in vessels built to Lloyd's requirements —and it cannot be urged that this society is too reckless in concessions of this nature—are 20 per cent. in scantling, and 18 per cent. in weight. As it is impossible to adjust the scantlings of material to take the full advantage of these reductions, and further, as allowance has to be made for extra weight due to the continued use of iron in vessels of steel— for purposes not essential to structural character — the average weight-saving effected in practice is about 13 to 14 per cent. This represents, in the finished vessel, a clear increase of at least 13 per cent. in dead-weight carrying power. The gain obtained in general practice has been otherwise stated on good authority as 7 to $7\frac{1}{2}$ per cent. of the gross tonnage.

In trades where there is constancy of dead-weight cargoes, this increase in dead-weight carrying power should speedily recoup the owners for extra first cost, and in the lifetime of vessels generally, a clear pecuniary gain should result. In trades, however, where the cargo consists of measurement goods, the advantages are not so decided, for it may sometimes happen that before vessels have been loaded to their maximum draught the limits of stowage will have been reached. Even here, however, the steel vessel has the advantage of her iron rival; her hull is 13 per cent. lighter, and consequently may be propelled at a given speed with much less expenditure of power, and has the further advantage—often a very important one—of a shallower draught. This latter consideration alone, in a service where every iota of such saving counts, has influenced many shipowners to adopt the steel.

As the manufacture of mild steel progresses and extends, the assimilation of the rival materials as to cost is sure to follow. Already very great advances have been made towards this end, the fact being abundantly evidenced by the greatly increased number of steel ships on hand, and by the establishment of new works, and transformation of old, for the better production of the new material. In 1877 mild steel was about twice as costly as the iron in common use. The sources of supply, however, were then comparatively few, and the thorough and severe testing to which the new material had to be subjected, necessarily increased the cost relatively to iron, which has never been subjected to the same rigorous ordeal. In 1880, owing to the increased sources of supply and the progress in manufacture, the cost of steel had been reduced, relatively to iron, by about 50 per cent. At the time of writing (March, 1883), the price of steel for a good-sized vessel is—overhead—about seven pounds, seven shillings and sixpence per ton; while the corresponding figure for iron is about five pounds, five shillings, or a difference of only about twenty-nine per cent. in favour of the older material.

Doubts were at first expressed by not a few, regarding the durability of steel ships compared with those of iron, such misgivings being aggravated by the thinness of the steel plating. This fear is being gradually lessened by the results of laboratory experiments and *bona fide* experience—the broad deduction from which is, that the deterioration of steel, under the action of sea water, is no greater than that of iron, and that, if the same care and constancy in cleaning and painting, common to ships of the latter material, be extended to ships of the former, their durability will be equal.

Several large shipowning companies were not slow to place faith in the new material. In the early part of 1879, the "Allan Line" Company entrusted to Messrs Denny & Brothers, of Dumbarton, the order for a huge vessel, which the intrepid confidence of the principal partners in both the owning and the building firms determined should be of mild steel, be bound with steel rivets, and have her boilers of the same material. This was the large steamer *Buenos Ayrean*, the first transatlantic steamer built with the new material. She was finished early in 1880, and had not been over nine months in

the water when the order for a second and still larger steel
vessel—the *Parisian*—had been given by the same owners to.
Clyde builders.   The Union Steamship Company of New
Zealand, the Pacific Steam Navigation Company, Messrs Donald
Currie & Co., and several smaller companies, ordered vessels
of steel almost simultaneously, while yet the new material was
in the early stage of trial.   Amongst the orders for steel
vessels which were subsequently given, the *Servia* and *Cata-
lonia*, for the Cunard Company; the *Clyde* and *Thames* and
Shannon for the Peninsular and Oriental Company; the *India,*
for the British-India Company; the *Arabic* and *Coptic*, for the
Oceanic Steam Navigation Company, and the four twin screw
steamers of the "Hill" Line, represent the principals.   The
companies who then adopted the new material have mostly
continued to have their new ships built of steel, and to name
the vessels since built and now building in which this material
is employed, would simply be to enumerate three-fourths the
fleet of high-class modern merchant ships.   There were 21,000
tons of steel shipping built throughout the United Kingdom
in 1879; 36,000 in 1880; 55,000 in 1881; 126,000 in 1882;
and over 244,000 in 1883.   It is computed that at the present
time the amount of steel shipbuilding going on throughout
the kingdom is not less than 175,000 tons, or the largest
amount on hand at any one time since its introduction.

The modification in the structural arrangement of ocean
trading vessels, already spoken of as the continuous-cellular
system, although only within very recent times receiving
extended adoption in the mercantile marine, possesses in some
of its essential features the prestige of years.   So long ago as
1854, Mr Scott Russell strongly advocated the principle of
longitudinal construction, and applied it in practice to ships
of the mercantile marine, to the success of which, in a scientific
sense, the *Great Eastern* is surely overwhelming testimony.
The principle met with much scientific favour from many
besides Mr Russell, but it did not take root in solid practice.

PORTRAIT
AND
BIOGRAPHICAL NOTE

———

NATHANIEL DUNLOP.

NATHANIEL DUNLOP.

BORN at Campbeltown, Argyleshire, in 1830, and educated at the Grammar School of that town. In 1845 removed to Glasgow, and in 1847 entered the counting-house of Mr George Gillespie, where he was chiefly employed in connection with the Allan Line service of clipper ships between Glasgow and Canada, for which trade Mr Gillespie was then agent. In 1853 transferred his services to the Allan Line firm, where, for several years, was principal clerk and cashier, subsequently becoming partner. During the year 1853 the Messrs Allan resolved to add a fleet of steamers to their already well-known line of clipper ships, and contracted for the building of four screw vessels, the first of which—the *Canadian*—was launched in July, 1854. The growth of the business may be inferred from the fact that the Allan fleet at the present time consists of twenty-eight steamers, of 87,078 tons, and fifteen sailing vessels, of 21,225 tons. Mr DUNLOP, since joining the firm, has taken an active part, along with Mr Alexander Allan, its senior member, in the building arrangements of the Allan Line. When mild steel was beginning to take the place of iron in the construction of steamers, and before any of the Atlantic companies had ventured on its use, Mr DUNLOP and his partners evinced ready confidence in the new material, their adoption of it being elsewhere referred to in this work. From an early period Mr DUNLOP has taken an active interest in shipping legislation. In 1874 gave evidence before the Select Committee of the House of Commons upon the Measurement of Tonnage Bill, and again in 1882 before the Royal Commission on the same subject. During the Plimsoll agitation, and the consideration of the proposed legislation resulting from it, was a witness before the Select Committee of the House. In 1879 was deputed by the Shipowners Association of Glasgow to give evidence before the Select Committee upon the Merchant Seamen Bill then before the House. In connection with Mr Chamberlain's recent efforts at legislation on Merchant Shipping, issued a pamphlet which very fully discussed the questions raised, and exhibited an analysis of the losses of life in merchant shipping. Gave evidence during the present year before the Load Line Committee, on which body Mr DUNLOP had been invited to serve ; business duties, however, preventing him accepting.

Yours faithfully,
Nath. Dunlop

Pecuniary and other kinds of considerations interposed to prevent its general adoption. The urgency for increase in the size of vessels was not such as to make longitudinal strength (the special advantage claimed for the new principle) a great desideratum; and there was perhaps reluctance on the part of shipbuilders to relinquish time-tried and familiar methods. The system presently under notice—although, as has already been said, the same, in its main principles, as the system then advocated—by its descent through the Admiralty Dockyards, by its application to merchant vessels—first of East Coast, and then of Clyde build—and by its close association with water ballast, has undergone many modifications which almost constitute it a creation of recent times.

Sir Edward J. Reed, when Chief Constructor of the Navy, introduced the bracket frame system of construction into iron-clad ships of war, and, as already indicated, it is largely owing to the experience of the system as applied and practised in such cases—conjointly, of course, with its successful introduction in the case of the *Great Eastern*—that in so short a time it has reached the present structural perfection, and received such wide extension in merchant steamships. That it has recently received such wide adoption in the mercantile marine is due not so much to its structural advantages—and these are great—as to the way in which it lends itself to the economical working of steamships in actual service. This will be more explicitly referred to after some description of the system as applied in merchant ships has been given.

It is somewhat away from the field this work is concerned with, to trace the system in its stages of development in ships of war, but it may be said, shortly, that the impulse which the system has received in the mercantile marine has in no sense been a transference of the activity which at all times since its introduction has characterised the application of the system to the vessels built in our naval yards.

In order to assist the non-technical reader in appreciating what follows regarding the system in merchant ships, a general idea of the cellular bottom principle of construction is afforded by Fig. 1.

FIG. I.

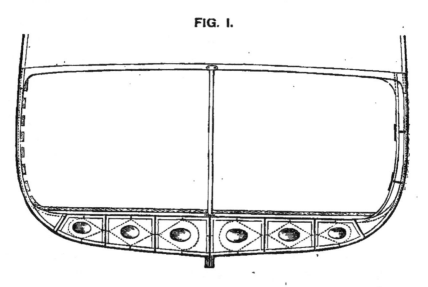

This shows in section the bottom part of a vessel amidships, fitted with a double or inner skin, extending across the ship from bilge to bilge, and there connected in a watertight manner to the outer bottom plating. A series of longitudinal plates are worked, fore and aft; set vertically between the outer skin of the vessel and the plating of the inner bottom, and connected thereto by continuous angles. Between these "longitudinals," and at every alternate transverse frame, deep plate floors, lightened with oval holes, are fitted, connected to outer skin by the angle frame, and to inner bottom plating by pieces of angles corresponding to the vessel's "reverse frames." These floor plates are, in addition, connected by vertical angles to the longitudinals. Intermediate between the deep plate floors simple angle bar transverse frames and reverse frames are fitted, to give support to the outer skin and to the inner bottom respectively. Until recently, the deep floors consisted of "gusset" or "bracket" plates, each division being fitted in four separate pieces, the whole taking the form as shown in dotted outline. This practice is still most largely followed, but in those yards which are equipped with large hydraulic punching machines for piercing holes such as are shown in Fig. 1, the solid floors have superseded the bracket or four-piece floors, the change effecting a simplification of work and decided structural advantages.

With the employment of water as a substitute for dry or rubble ballast, the structural movement under notice may be said primarily to have begun. This movement has resulted in the present approved system, which, at the same time that it has regard to water-ballast with all its attendant advantages, most happily combines the important qualities of increased strength and security. The need for ballast in vessels whose service generally comprises " light " as well as " loaded " runs (as in the coal trade between Newcastle and London), or in trades where the full complement can only be obtained by shifting from port to port, is obviously great. It is doubtless to needs such as these, more than to any demand for increased structural strength, that the introduction and extended application of the longitudinal and bracket-plate principle is owing.

The screw-steamer *Sentinel*, built in 1860 by Messrs Palmer of Jarrow, Newcastle-on-Tyne, is mentioned by some authorities as embodying some of the main features of the longitudinal and cellular bottom system, and the screw-steamers *Scio* and *Assyria*, of 1440 tons, built in 1874 by Messrs Westerman, near Genoa, have been noticed in a similar connection. The next vessel, in point of time, which contained features answering to the system now in vogue, and from the date of whose production the movement has been almost constantly progressive, was the screw-steamer *Fenton*, built by Messrs Austin & Hunter, of Sunderland, in 1876.

Clyde builders were not slow to recognise the value of the system in its application to water-ballast steamers, and almost immediately some of the more intrepid of their number began to advocate its adoption, but with some modifications, in vessels then being contracted for. Mr John Inglis, jun,. of Messrs A. & J. Inglis, Pointhouse, Glasgow, submitted to Lloyd's Registry in March, 1878, the scantling section of some cellular bottom vessels, then in project, which contained several of the

improvements introduced in subsequent practice.   Messrs
William Denny & Brothers, of Dumbarton, at the same time
took up the principle, and have since actively applied it to
steamers of every character in which water-ballast is a
desideratum.   Adopting it, five years ago, in four sister vessels
for the British India Steam Navigation Coy., they subsequently
raised the important issue with the Board of Trade regarding
the tonnage measurement of these vessels.   This august body
insisted on computing the register tonnage—the figure upon
which the tonnage dues are levied—not to the top of the inner
bottom, but to an imaginary line half-way down the cellular
space—in fact, to where the line of floor would have been if
constructed in the ordinary fashion.   Messrs Denny maintained, in effect, that as the register tonnage was meant to be a

FIG. 2.

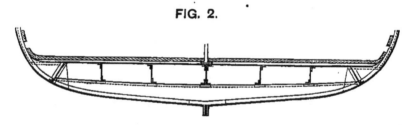

measure of the space available for cargo, the top of the ceiling
on the inner bottom was the only equitable line of measurement.   The principal reason for the Board seeking to pursue
this course seems to have lain in the supposition that owners
would endeavour to use the double bottom for cargo-carrrying
purposes.   An ambiguity in the words of the Merchant
Shipping Act, or their inapplicability to present day practice,
were other possible elements in the case, but doubtless the
red-tapeism and self-sufficiency characteristic of the Board
had much to do with their action.   This is borne out by the
fact that although the Messrs Denny succeeded in their plea
with respect to vessels having structural cellular bottoms, the
absurd practice is still followed in cases where the bottom
is fitted for water ballast on the girder principle, *i.e.*—the
inner bottom fitted upon fore and aft runners or girders,
erected on floors of the ordinary description, as shown in Fig. 2.

This formed, and still forms in many places, a very common arrangement for water ballast steamers, although not so inherent a feature of the vessel's structure as the continuous cellular bottom. In most cases this system is fitted only for part of the length, and not, like the cellular system, applied throughout the whole length of the ship. If it was impossible for the Board of Trade to hold by the contention that cargo might be carried in bottoms of the structural cellular type, it is equally untenable in the case of bottoms such as are now referred to. The difference between the two kinds of ballast bottoms is one merely of construction, and if any one of the two lends itself to cargo-carrying purposes, it is certainly the cellular system. The anomaly is sufficiently striking to merit attention, and in certain districts where the girder system is largely adopted for medium-sized vessels, it is felt as nothing short of an injustice, both by shipowners and builders.

The concession or victory won by Messrs Denny removed a serious hindrance to the spread and general adoption of the water ballast cellular system. Other Clyde firms at the same time—or at least soon after the adoption of the system by the Messrs Denny—took the matter up and independently did much towards the popularisation of the cellular mode of construction. Speaking in the early part of 1880, Mr William John, of Lloyd's Registry, now General Manager with the Barrow Shipbuilding Company, said:—"At the time Mr Martell read his paper on water-ballast steamers before the autumn meeting of this Institution (Naval Architects) at Glasgow, in 1877, there had been only two or three small steamers built (since Mr Scott Russell's early ones) on the longitudinal principle. Now, it is within the mark to say there are one hundred steamers, built and building, whose bottoms are constructed on the longitudinal principle, or what is better described as the cellular system, amounting probably to 200,000 tons, and it is not outside the bounds of probability that a very few years will see the majority of merchant steamers constructed in this manner." Mr John's connection with Lloyd's at the time, entitled his statements and opinions

with regard to the prevalency and prospects of cellular construction to be accepted with every assurance, for it is in such
Societies as Lloyd's where the best concensus of information
regarding the extent and tendencies of particular types of
vessels can be obtained. In point of fact, the intervening
period has witnessed, in great measure, a realisation of Mr
John's forecast. The advantages of a cellular bottom as
regards safety, and for the purpose of ballasting and trimming
vessels, also as meeting the greater need for longitudinal
strength caused by the enormous growth in the size of vessels,
have received that appreciation from shipowners and shipbuilders which is their due. The practice has accordingly
spread, till now, it would not be rash to say, quite as many
of the ocean trading steamers being built are fitted with
cellular bottoms as are without them.

The adaptation of water ballast to sailing vessels, as well as
to steamers, has received consideration at the hands of both
Tyne and Clyde builders. Previous to 1877, several small
sailing ships were built on the Tyne, in which provision was
made for water.ballast in tanks entering into the structure of
the bottom, but erected over the ordinary plate floors. About
150 tons of water ballast were carried by these vessels, the
filling and discharge of the tanks being effected by Downton's
pumps, worked by the crew. The trade in which they were
engaged—*i.e.*—carrying coal from the Tyne to Spanish ports,
and back to this country with ore—was one in which the
introduction of water ballast proved commercially and otherwise most advantageous. Two years subsequently Messrs A.
M'Millan & Son, Dumbarton, introduced water ballast into
one of the largest class of sailing vessels then being built.
Unlike previous sailing ships with provision for water ballast,
however, the vessel was constructed on the structural cellular
bottom principle, having bracket floors and continuous girders,
as so generally approved in steamships. Capacity for water
ballast, to the extent of over 300 tons was thus provided, the
filling and discharge being effected by a special donkey engine,
supplied with steam from a large donkey boiler. The boiler

also furnished the motive power for cargo winches, off which, by crank gear, the manual labour pumps were also brought into requisition. Facilities for the expeditious management of ballast—the want of which, in sailing vessels, considerably hinders its adoption—were thus, in this case, efficiently provided. Several other sailing ships, built by Messrs A. M'Millan & Son, and by other shipbuilding firms on the Clyde, have been fitted with this system, and the result of experience with these vessels in actual service, thoroughly encourages its more general adoption.

Many minor, yet aggregately important, structural features which are products of the progressive movement of recent years, or are simply revivals of old devices which were "untimely born," still call for some notice. As a necessary consequence of the growth in dimensions and the change in relative proportions of vessels, greater regard has been paid to the systems of construction in which the longitudinal principle is involved. This, of course, is evidenced by what has been said of the cellular bottom system, but various minor structural features associated with the cellular bottom are also noteworthy in this connection. It is the practice, for instance, where large ships are concerned, to fit side stringers in the holds, throughout the entire length, made intercostal with regard to transverse plate or web-frames occurring at intervals of 16 or 20 feet, which extend from the bilge to the main deck. This arrangement—an outline of which may be found to the right of the section shown as Fig. 1—possesses many structural advantages, and finds additional favour with shipowners on account of its leaving a clearer hold for stowage by obviating the use of transverse hold beams.

Regard for transverse strength has increasingly evinced itself in the fitting of various kinds of plate side stiffeners or partial bulkheads. This is well exemplified in a very recent case— that of the National Company's steamship *America*, built by Messrs J. & G. Thomson. This vessel, having been constructed independent of any special Registry Rules, embodies structural features not common amongst vessels in which such rules are undeviatingly conformed to. The system referred to, of plate frames or partial bulkheads, is one of the most conspicuous of these features. Throughout the length of the vessel, at intervals of about 18 feet, transverse plate stiffeners or frames, extending from the shell inwards about 4 feet, take the place of the ordinary angle frames, and are continuous from floors to upper deck, the stringers and other longtitudinal features being scored through them. The surplus transverse strength resulting from this system is such as amply to compensate for uncommonly large breaches made in the deck beams and plating for light and air purposes in the saloons. This is a very special feature in the interior arrangement of the *America*, and will be referred to further on. The regard for transverse strength, again, conjointly with the increased attention to minute watertight sub-division, has led to the fitting of a greater number of complete watertight transverse bulkheads, relatively to the lengths of vessels.

In vessels of extreme proportions the method of forming shells two-ply, or of fitting all the shell plates edge to edge with outside covering-strakes over the fore-and-aft joints, has been recently revived and much improved. The system, although very expensive, has been adopted in vessels for the Anchor Line by Messrs D. & W. Henderson, Glasgow, and subsequently on even a more extensive scale by the Barrow Shipbuilding Company.

Affecting the structural character of modern ships very materially, but the result chiefly of an economy in labour, rivetting by machine power has received a wonderfully extended application within recent years. Structurally, as well as commercially, the system has played a large part in the progressive movement under review. By its means the strength of united parts has been enhanced through the increase of their frictional resistance, and through the rigidity

PORTRAIT
AND
BIOGRAPHICAL NOTE
——
THOMAS HENDERSON.

THOMAS HENDERSON,

CHAIRMAN OF THE GLASGOW SHIPOWNERS' ASSOCIATION ; OF THE LOCAL MARINE BOARD OF THE PORT OF GLASGOW AND OF THE CLYDE LIGHTHOUSE TRUST ; DIRECTOR OF THE GLASGOW CHAMBER OF COMMERCE, AND OF THE CHAMBER OF SHIPPING OF THE UNITED KINGDOM.

MR THOMAS HENDERSON, senior member of the firm of Henderson Bros., managing owners of the Anchor Line of Steamships, is a native of Fifeshire, but was educated in Glasgow. He entered, at an early age, the mercantile marine service as an apprentice, and rapidly rose through the different grades of the profession to the command of various sailing ships and steamers belonging to the port of Glasgow. In 1853 he was admitted a partner in the shipping firm of Handyside, & Co., which, five years afterwards, was changed to Handyside & Henderson. Some years later, on the retirement of the Messrs Handyside and the assumption of Mr John Henderson and other partners into the business, the firm became Henderson Brothers, under which designation the greater part of the steam shipping business now carried on by the Anchor Line steamships has been developed and extended. The fleet as now constituted consists of forty-five steamships of an aggregate measurement of over 124,000 tons, with an engine power of above 25,000 horses nominal. These vessels are employed severally in the Transatlantic, Indian, and Mediterranean services, in all of which they are well known and appreciated by the public as in all respects first-class, and second to no other competing line for safety, speed, comfort to passengers, and careful delivery of goods carried. One branch of the extensive services of the Anchor steam ships, specially noteworthy as forming one of the modern "express" lines which have given such impetus to ocean travel, is the express service between Liverpool and New York, in which the magnificent steamships *City of Rome* and *Austral* are engaged. In connection with their head office in Glasgow, Messrs Henderson Bros. have established branch offices of their own in London, Liverpool, Manchester, Barrow-in-Furness, Queenstown, Londonderry, Dundee, New York, Boston, Chicago, Paris, Marseilles, and Palermo, at all of which the agency business of the several lines of steamers is attended to by their own employees. In addition to his responsible share in the concerns of the Anchor Line, Mr HENDERSON is a partner in the extensive shipbuilding and engineering works of D. & W. Henderson & Co., at Meadowside, Partick, and Finnieston Quay, Glasgow. The estimation in which Mr HENDERSON is held as a shipping and commercial authority may be inferred from the enumeration of important offices at the head of this note ; most of which he has worthily occupied for many years.

Very truly yours
Thomas Henderson

of joints, due to the more thorough filling of the rivet holes. The subject of hydraulic or machine power rivetting will, however, receive fuller treatment in a subsequent chapter.

Within the past two or three years cast steel stems, stern frames, and rudders, have been taking the place of forged iron work in ship construction. The practicability of manufacturing these of such strength and homogeneity as would meet the needs of ship construction even better than the ordinary forged work, had occurred some five or six years ago to several engaged in the steel trade. Mr. J. F. Hall, of Messrs William Jessop & Sons, Limited, Sheffield, had the subject under consideration about that period, and actually made several small stern posts and rudders for steam yachts and launches. The advantages of solid and uniform steel castings over iron forgings—which, with their many weldings, so often prove inefficient when subject to any sudden shock—were even then rightly enough appreciated. It was only, however, after patents had been taken out by Messrs Cooke & Mylchreest, of Liverpool, for various devices connected with the actual fitting of such features to the ship's structure—amongst other things the hanging of rudders without pintles or gudgeons—that the manufacture of cast steel stern-frames, rudders, &c., was seriously proceeded with.

In July, 1882, the Steel Company of Scotland (Limited), who are the manufacturers in Scotland of Messrs Cooke & Mylchreest's patent form of rudders and stern-frames, successfully cast a stern-frame—the first of large size, it is believed, made for actual use in the construction of a steamer. In April of the same year, however, Messrs William Jessop & Sons (Limited), of Sheffield, had exhibited a crucible cast steel stern-frame and rudder of their manufacture, at the Naval and Sub-Marine Exhibition, held in London. These large castings, along with others, were subjected to a series of tests in the presence of Lloyd's inspectors and other authorities, such as the forged frames and rudders ordinarily fitted would not have come through without severe damage, yet all of which the steel castings withstood most thoroughly.

Testimony to the efficiency of these new features in ship construction has already been furnished from the arena of actual experience, by the recent grounding of two steamers in which these features had been introduced. The screw-steamer *Euripides*, a Liverpool-owned vessel of about 1780 tons gross, completed in May, 1883, by Messrs Caird & Purdie, of Barrow, some time ago ran upon a reef of boulders, and remained thumping heavily for several hours. At the time she was laden with a full cargo of grain, which was afterwards delivered in perfect condition. The cast steel stem and stern-frame, which were manufactured by the Steel Company of Scotland, were practically without damage, notwithstanding that serious indentations were made in them. The stem, although receiving the full force of resistance, was not perceptibly altered in shape, and competent judges who inspected the damage in dock were of opinion that the stem, with its superior attachments, in all probability saved the vessel from total loss. The rudder on the *Euripides* is of solid cast steel, in one piece, and hung without pintles, and in a manner involving little or no rivetting. In this, as in the other features, the immunity from serious damage testifies to the efficiency and durability of the steel castings. The second case of grounding referred to is that of the screw-steamer *Strathnairn*, of 400 tons, belonging to Messrs James Hay & Sons, of Glasgow; one of two vessels built by Messrs Burrell & Son, of Dumbarton, in which cast steel stern-frames and rudders were adopted. This vessel got aground while off Harrington, on the north-west coast of England, about the latter end of March of the present year. Her stern-frame sustained very considerable shock: such, indeed, as no ordinary forged work could possibly have undergone with like result. Subsequent docking showed that it would only be necessary to straighten the frame at the deflected portions in order to make it again structurally efficient. This was done, and the vessel is again actively engaged in service.

The weldless stern-frames, rudders, and stems, as patented by Messrs Cooke & Mylchreest, Liverpool, and manufactured for them by the Steel Company of Scotland, Messrs Jessop.& Sons, Sheffield, and Messrs John Spencer & Sons, Newcastle, have various advantageous features which may be noticed somewhat fully. One of these is the casting of flanges on the sternposts, for attaching the shell plates to; by which arrangement much of the difficult and costly work in the rivetting and fitting of the shell plates at these parts is done away with, while a considerable increase of strength is obtained. The solid rudder, is a great improvement on the built rudder as usually fitted; the entire absence of rivets being an important desideratum. The rivets connecting the rudder-plates to the frame-forging are frequently a source of trouble and annoyance, through their being loosened by the constant vibration of the rudder, and the shocks it often receives. The heads of the rivets not unfrequently drop off, and the rivets themselves sometimes fall completely out. All this, of course, is entirely obviated in the solid rudder. By Messrs Cooke & Mylchreest's improved method of fitting the rudder—a device which is only applicable in a casting—pintles are wholly dispensed with, and in their place a much stronger joint is substituted, with a considerably increased wearing surface. The rudder is also jointed at the top of the blade, by means of strong flanges bolted together; an obvious advantage of this arrangement being that it can be readily unshipped, even when afloat.

In addition to the stern-frames, stems, and rudders, there are, also being supplied, keels, garboard strakes, and centre keelsons in long lengths. It is claimed for these that as the keel, garboard strake, keelson, and brackets for connecting the floors, are all made in one piece, they are much stronger than as ordinarily constructed, and that a considerable saving in both labour and rivets is effected. As there are no angle irons to contend with, the limber-holes may be made close to the bottom plating, and a much thinner layer of cement will, consequently, be needed on the bottom; the saving in this respect, according to the patentees' calculation, being 50 tons in a 2,000-ton vessel.

As the prices of these frames and rudders do not exceed those charged for frames of wrought-iron, and moreover, owing to the pieces which are cast on to them forming attachments for keels, decks, &c.—thus cheapening the work of construction in the shipyard—there appears to be no question of their great superiority. The presence of blow-holes, not unfrequently a source of misgiving in castings, is found from experience to be a constantly diminishing fault in these articles. The demand for them has steadily grown since their adoption in a few actual cases. It would seem, indeed, that the demand is only limited by the powers of production possessed at present by the four or five steel-making firms who have undertaken this class of work, and have satisfied the requirements of the registration and the insurance societies.

In addition to the frames and rudders for ordinary screw vessels, the Steel Company of Scotland have also supplied several stems for war vessels, with rams and torpedo openings, which have proved very satisfactory. Other new adaptations are the casting of large brackets for shafts of twin screw vessels, of large crank shafts themselves, and of heavy anchors; the results of tests presently being made fully warranting the anticipation that the material will very largely be employed in the future for these important items in the outfit of merchant vessels.

The more important features of growth or change in ship construction which have made the past few years a noteworthy period in the history of mercantile shipbuilding have now been reviewed. Speed, and propulsive power of steamships, although absorbing very much of the progress for which the period has been so remarkable, have not been dealt with, but are reserved for the chapter following. The subjects named will also necessarily receive some attention in the chapter devoted to progress in the science of shipbuilding. In anticipation, however, apologies should be offered for the paucity of detailed references to the propulsive agents on board ship.

Marine engineering, in all its recent developments, would require for its proper treatment considerably more space than can be devoted to it in the present work.

To meet the exigencies of the progressive movement, both practical skill, scientific knowledge, and commercial enterprise have been needed on the part of our shipbuilders. These have not been by any means wanting, as abundantly evidenced by the foregoing record of what has been achieved. With a continuance of that readiness displayed by shipbuilders and naval architects to modify, and even revolutionize if need be, types and methods which the times have outgrown, the lead in merchant shipbuilding will long be ours. With a maintenance also of the enterprise shown by our shipowners, Britain will still continue, as regards the number, size, and power of her merchant ships, supreme among the nations.

List of Papers and Lectures bearing on recent improvements in ship design and construction, to which readers desiring fuller acquaintance with the *technique* and details of the subjects are referred:—

ON A NEW MODE OF CONSTRUCTING IRON AND COMPOSITE SHIPS, by Mr J. E. Scott: Trans. Inst. Engineers and Shipbuilders, vol. xv., 1871-2.

ON THE STRENGTH OF IRON SHIPS, by Mr William John: Trans. Inst. N.A., vol. xv., 1874.

ON TRANSVERSE AND OTHER STRAINS OF SHIPS, by Mr William John: Trans. Inst., N.A., vol. xviii., 1877.

ON A SYSTEM OF SHIPBUILDING COMBINING TRANSVERSE AND LONGITUDINAL FRAMING, by Mr James Hamilton, Jun.: Trans. Inst. Engineers and Shipbuilders, vol. xviii., 1877-88.

ON THE LONGITUDINAL BULKHEAD SYSTEM OF IRON SHIP CONSTRUCTION, by Mr Edwin W. De Rasset. Trans. Inst. N.A., vol. xviii., 1876.

ON IRON AND STEEL FOR SHIPBUILDING, by Mr Nathaniel Barnaby: Trans. Inst. N.A., vol. xvi., 1875.

ON STEEL FOR SHIPBUILDING, by Mr Benjamin Martell: Trans. Inst., N.A., vol. xix., 1875.

ON THE USE OF MILD STEEL FOR SHIPBUILDING in the French Dockyards, by M. Marc Berrier-Eontaine : Trans. Inst., N.A., vol. xxii., 1881.

ON STEEL IN THE SHIPBUILDING YARD, by Mr William Denny : Trans. Inst. N.A., vol. xxi., 1880.

ON THE ECONOMICAL ADVANTAGES OF STEEL SHIPBUILDING, by Mr Wm. Denny : Journal (No. 1) Iron and Steel Institute, 1881.

ON IRON AND STEEL AS CONSTRUCTIVE MATERIALS FOR SHIPS, by Mr John Price. Proceedings Inst. Mech. Engineers, 1881.

ON STEEL, by Mr James Riley ; Lectures on Naval Architecture and Marine Engineering : Glasgow, William Collins & Sons, 1881.

ON WATER BALLAST, by Mr Benjamin Martell : Trans. Inst. N.A., vol. xviii., 1877.

ON THE CELLULAR CONSTRUCTION OF MERCHANT SHIPS, by Mr William John : Trans. Inst. N.A., vol. xxi., 1880.

ON THE INCREASED USE OF STEEL IN SHIPBUILDING AND MARINE ENGINEERING, by Mr John R. Ravenhill : Trans. Inst. N.A., vol. xxii., 1881.

ON THE STRUCTURAL ARRANGEMENTS AND PROPORTIONS OF H.M.S. "IRIS," by Mr W. H. White : Trans. Inst. N.A., vol. xx., 1879.

ON THE QUALITY OF MATERIALS USED IN SHIPBUILDING, by Mr H. H. West : Trans. Inst. N.A., vol. xxiii., 1882.

ON THE USE OF STEEL CASTINGS IN LIEU OF IRON AND STEEL FORGINGS FOR SHIP AND MARINE ENGINE CONSTRUCTION, by Mr William Parker : Journal, Iron and Steel Institute, 1883.

SOME CONSIDERATIONS RESPECTING THE RIVETTING OF IRON SHIPS, by Mr Henry H. West : Trans. Inst. N.A., vol. xxv., 1884.

RECENT IMPROVEMENTS IN IRON AND STEEL SHIPBUILDING, by Mr William John : Iron and Steel Institute, 1884.

CHAPTER II.

IN these days of feverish activity in every avenue of business, when even leisure has come to be observed at a much more accelerated *tempo* than formerly, speed in locomotion would seem to be the first desideratum, not only on shore but afloat as well. In no ocean service is the truth of this so apparent as in the transatlantic mail and passenger service, the oldest and most constantly progressive, and where at the present time, certainly more than at any former period, the contest for supremacy amongst rival steamship lines has assumed the form of increased speed and enhanced passenger accommodation.

The Atlantic service, for these reasons, as well as because it exemplifies more of the fruits which have rewarded the joint labours of the engineer and shipbuilder in improving marine propulsion, may be selected for detailed review. In other ocean services, of course, the achievements of engineering and shipbuilding skill have also been made apparent, and in ways, perhaps, which the Atlantic service does not exhibit. Reference to these will afterwards be made, but attention will meantime be confined to the service stated, and to such considerations of the general progress made in ocean navigation as are necessarily involved in the particular subject.

It is needless, in view of the frequency with which the story of ocean steam navigation is told, and especially, considering the scope of the present review, to enter at any length into the details of early service. The first practically successful transatlantic steamers were the *Sirius* and the *Great Western*, the first a paddle steamer 170 feet long, 270 horse power

originally constructed to ply between London and Cork, and
the latter, a paddle steamer, 212 feet long and about 440
horse-power, designed and built expressly for the transatlantic
service.   The *Sirius* left Cork on the 4th April, 1838, and
reached New York on the 22nd; the *Great Western* left
Bristol on the 7th April, three days after the *Sirius*, reaching
New York on the 23rd—the time taken being thus 18 days
and 15 days respectively.   The return voyages of these pioneer
long-passage steamers were made in 16 days and 14 days
respectively, their performances at once establishing the
superiority of steamers, commercially and otherwise, over the
sailing ships which had previously for so long been the recognised medium of transit in the Atlantic passenger trade.

In 1840 a regular mail service by steamers was first introduced on the Atlantic.   The first of these mail steamers was
the Cunard paddle-steamer *Britannia*, 207 feet long. which
sailed from Liverpool on July 4, 1840, and arrived at Halifax
in 12 days 10 hours, the return journey being performed in 10
days.   The *Acadia*, *Columbia*, and *Caledonia* all of about the
same dimensions as the *Britannia*, at once followed.   The
success of the Cunard Line was so marked that opposition was
soon provoked, and in 1850 the Collins Line of American
steamers started to compete with the Cunard liners.   The
same year also saw the commencement of the well-known
Inman Company, of Liverpool, their first vessel being the *City
of Glasgow*, an iron screw steamer of 1680 tons and 350 horse-
power.   The Allan and Anchor Lines were established in 1856,
the Guion Line in 1863, and the White Star Line in 1870.

With the substitution of the screw propeller for the paddle
wheel, first carried out to any great purpose in the small
steamer *Archimedes* in 1839, but introduced with even greater
effect in the Atlantic steamer *Great Britain* in 1843, was laid
the basis of that progressive and magnificent success in propulsion which has since attended ocean navigation.   It was
with screw-steamers Mr. Inman boldly assailed the Cunard
Company in 1850, but notwithstanding this, it was only in
1862 that the Government consented to sanction the use of

the screw in the mail steamers of the Cunard Company. The *Scotia*, measuring 366 feet in length, by 47½ feet in breadth, and 30½ feet depth, launched in 1861, was the last paddle steamer built for this company.

The other great improvements contributing to the success spoken of, were the introduction of engines designed on the compound principle, and a little later, the employment of the surface condenser, and the use of circular multitubular boilers. In spite of the success with which the compound system was attended in vessels built for the Pacific Steam Navigation Company as early as 1856, and for some other private owners soon after, the great steamship companies, and shipowners generally, were very slow to adopt it. It was not until about the year 1869 that the compound engine came into general use, and it was only in 1872 that the Cunard Company seriously took it into favour.

The early steamers of the Cunard Line possessed an average speed of 8½ knots, and took about 15 days for the voyage Through the Collins rivalry the speed was increased to an average of 12½ knots, and the time for crossing the Atlantic was reduced to 12 days 9 hours outwards, and 11 days 11 hours homewards. In 1856, the powerful paddle-steamer *Persia* (the first iron vessel built for the Cunard Company) was placed on the service, and attained an average speed of about 13 knots, consuming 150 tons of coal per day. She made the distance between Queenstown and New York, on an average, in 10½ days. In 1862 the *Scotia*, belonging to the same company, made the passage in 9 days.

Coming down to more recent times, the White Star Line, with its steamships *Britannic* and *Germanic*, built in 1874 and 1875 respectively, held for a considerable period first place in the matter of fast steamships. The vessels named were, however, in time beaten by the newer ships *Gallia*, of the Cunard Line, and *Arizona*, of the Guion Line. As illustrating the speed at which the vessels named accomplished the transatlantic voyage—between Queenstown and New York— the following brief list, compiled from published records, of fast

runs out and home during the period 1875-1881, may here
be given :—

Vessels.	Out.			Home.		
	Date.	Time.		Date.	Time.	
		D. H. M.			D. H. M.	
Britannic,...............	Aug.,1877,	7 10 50		——		
Britannic,...............	May, 1879,	7 13 7		May, 1880,	7 19 22	
Germanic,.........	Oct., 1880,	7 13 0		Nov., 1881,	7 17 34	
City of Berlin,.........	Oct., 1877,	7 14 12		Oct., 1875,	7 15 48	
City of Berlin,.........	Oct., 1880,	7 20 32		Sep., 1879,	7 19 23	
City of Richmond, ...	Oct., 1880,	8 0 0		July, 1879,	8 3 52	
Gallia,....................	May, 1879,	7 22 50		May, 1881,	7 18 50	
Arizona,.,...............	Sep., 1881,	7 8 32		Sep., 1881,	7 7 48	

When the success of vessels of the size of the *Arizona* and
the *Gallia* was made apparent, it was decided by the Cunard
Company to build a larger and faster ship than previous ones.
Accordingly, in the autumn of 1880, specifications were issued
to some of the leading shipbuilding firms, asking them to
tender for the construction of a vessel of 500 feet in length,
50 feet beam, and 40 feet depth. At the suggestion of Messrs
J. & G. Thomson, who were successful in securing the contract
for this remarkable vessel, the dimensions were increased to
530 feet by 52 feet by 44 feet 9 inches. With these dimensions, and with mild steel as the constructive material, the
new vessel—the *Servia*—was thereafter proceeded with in
Messrs Thomson's establishment.

The Guion line, not to be left behind, placed the order for a
vessel of the dimensions first proposed for the *Servia*, with
Messrs John Elder & Co., but, in order to be faster than the
Servia, the weight-carrying was considerably reduced, and the
boiler power much increased. The wisdom of this step has
been justified by the now generally received opinion that these
fast steamers should not carry such heavy cargoes as the slower
ones. This new vessel for the Guion line was the *Alaska*, now
justly noted for her fast runs across the Atlantic.

The Inman Company also decided not to lag behind, and
as soon as the conditions of the design of the *Servia* had been
fixed, they placed the order for a ship—the *City of Rome*—with

PORTRAIT
AND
BIOGRAPHICAL NOTE.

———

WILLIAM PEARCE.

WILLIAM PEARCE,

MEMBER OF COUNCIL OF THE INSTITUTION OF NAVAL ARCHITECTS;
MEMBER OF THE IRON AND STEEL INSTITUTE, AND OF THE INSTITUTION
OF ENGINEERS AND SHIPBUILDERS IN SCOTLAND.

———

BORN at Brompton, in Kent, in the year 1835. Learned practical shipbuilding in Her Majesty's Dockyard at Chatham, and was at the same time engaged in the office of the master shipwright there, the late celebrated Mr Oliver Lang. When the Government in 1861 determined upon the construction of iron ships in the Royal Dockyards, was the first officer selected to carry on that work, and superintended the building of H. M. *Achilles* in the dockyard at Chatham. In 1863 left the Government service to become a Surveyor to Lloyd's Registry in the Clyde district, and in 1864 was appointed General Manager in Messrs R. Napier & Sons' shipbuilding establishment, where, in 1865, his ability as a naval architect was first brought into prominence through the designing of the *Pereire* and *Ville De Paris*, built for the Compagnie General Transatlantique, which vessels maintained for several years a foremost place amongst the fast ships on the Atlantic. After the death of Mr John Elder, in 1869, joined by request the late Messrs John Ure and J. L. K. Jamieson in carrying on and extending the gigantic shipbuilding and engineering business at Fairfield, under the title of John Elder & Co. In 1878 Mr Ure and Mr Jamieson retired from the firm, and Mr PEARCE became sole partner, which position he has occupied up to the present time. Has constructed many steamships that are amongst the most celebrated in existence, of which it may suffice simply to name the *Arizona*, *Alaska*, and *Oregon*; the *Orient*, *Austral*, and *Stirling Castle*; also the *Umbria* and *Etruria*, just being completed for the Cunard Steamship Company. Another vessel built by Mr PEARCE, the construction of which excited, perhaps, a greater amount of interest than any of the above-named, was the yacht *Livadia*, for the late Emperor of Russia. The design, which was a fantastic one, was by Admiral Popoff. Mr PEARCE's enterprize has not been confined to shipbuilding and engineering, having projected or become largely interested in several lines of steamers, amongst which are, the Pacific Mail Steamship Co.; the New Zealand Shipping Company; the Guion Line; and the China Line of the Scottish Oriental Steamship Company. In 1880 Mr PEARCE gave the opening lecture in the course delivered in connection with the Marine Exhibition held in the Corporation Buildings, Glasgow. In 1881 was appointed a member of the Royal Commission on Tonnage, and in October of the present year was appointed a member of the Royal Commission on Merchant Shipping.

the Barrow Shipbuilding Company, intended to be larger, finer, and faster. Expectations as to speed and carrying powers were not in her case fulfilled, and the result of the dissatisfaction which this occasioned, was, that the *City of Rome* changed ownership, Messrs Henderson Brothers, of Anchor Line fame, coming into possession. In the hands of its new owners, the *City of Rome* was re-arranged internally, and her boiler power was considerably augmented, while her engines also were thoroughly revised. When first built, the vessel was fitted with engines of 8500 horse-power. As revised, they indicate 12,000 the acquisition being largely due to the fitting of four additional boilers. The results which have accrued from the extensive alterations made are such as to have firmly established the vessel in a foremost place in the Atlantic service.

The performances of the vessels named have been the subject of considerable interest to all concerned in shipping affairs, and to the public generally. The following table of fast passages accomplished during the past two years by these vessels has been compiled from published records, and from information supplied by the shipowning companies :—

	Out.				Home.			
Names of Vessels.	Date.	Time.			Date.	Time.		
		D.	H.	M.		D.	H.	M.
Alaska, ...	April,1882,	7	4	32	June,1882,	6	22	0
Do., ...	May, 1882,	7	7	0	Sep., 1882,	6	21	48
Do., ...	May, 1882,	7	4	10	Jan., 1883,	6	23	42
Servia, ...	Jan., 1882,	7	8	13	——			
Do., ...	Aug.,1883,	7	6	0	——			
City of Rome, ...	May, 1883,	7	12	16	June,1883,	7	7	4
Do., ...	June,1883,	7	4	56	July,1883,	7	2	19
Do., ...	Aug.,1883,	6	22	6	Aug.,1883,	6	21	4
Do., ...	Sep., 1883,	7	3	0	Sep., 1883,	6	23	24

An addition to the list of competitors was made in the *Aurania*, built by Messrs Thomson in 1882, and tried in June, 1883, when she attained a mean speed of 17¾ knots, and showed herself not unequal to a maximum speed of 18½ knots under circumstances ordinarily favourable. An untoward and serious accident to her machinery laid the *Aurania* aside just as her capabilities in actual service were being shown. It is during the "passenger season" that the qualities of these transatlantic steamers are best brought out, and it remains with the season which has just begun, to demonstrate to the full the *Aurania's* powers.

A similar remark applies to the *Oregon*, a still more recent competitor from the same stocks as the *Alaska*, whose dimensions correspond with those of the *Alaska*, except in respect to breadth, the first-named vessel having 3-ft. 6-in. more beam than the latter, the figures being—length over all, 520-ft.; breadth, 54-ft.; depth, 40-ft. 9-in. Extra power of engines to the extent of nearly 3000 horses indicated has been fitted in the *Oregon*. On the occasion of her speed trial on the Clyde she ran the distance between Ailsa Craig and Cumbrae Head —29½ nautical miles—in 1 hour 20 minutes, or about equal to 20 knots per hour. This was attained with the engines indicating 12,382 horse-power and making 62 revolutions per minute, the steam pressure being 110-lbs. per square inch. This result was doubtless attained under conditions more favourable to speed than the vessel is, as a rule, likely to meet with in actual service; and, as has been indicated, it still remains with the future to determine how far the aims of the owners and builders of the *Oregon* are realised.*

In the *America*, launched from the yard of Messrs J. & G. Thomson, near the close of 1883, and presently being fitted for sea, the National Steamship Company (Limited), of Liverpool, have embodied the results of their careful study of the development and changes in the mode of conducting the American trade. From such experiments—for they can hardly be con-

---

* Since the above was written, the *Aurania* and the *Oregon* have resumed their services on the Atlantic, the results in the case of the latter vessel being extraordinarily successful. On Saturday, the 5th April, she arrived at Queenstown, having left New York on Saturday, the 29th March, making the trip in 7 days, 2 hours, 18 minutes, her daily runs being :—45, 407, 396, 400, 302, 410, 384, 412, and 60; total, 2816 knots. Leaving Queenstown on Sunday, the 13th April, she arrived at New York on Saturday, the 19th April, in the unprecedentedly short period of 6 days, 9 hours, 22 minutes.

sidered anything else—as the rapid passages of the *Alaska*, the *City of Rome*, and other "greyhounds of the Atlantic," the company see it is no longer possible or profitable to have "composite" vessels—*i.e.*, those intended to carry a large cargo as well as passengers,—but that practically one class of vessels must be built for the passenger traffic and another for the conveyance of cargo. The vessel represents an attempt to solve the problem of producing a ship which shall have large passenger accommodation and a high speed, with a comparatively small first cost and a reasonable consumption of coal. She is built of steel, and of the following dimensions :— Length, 440 feet; breadth, 51¼ feet; depth of hold, 36 feet; gross tonnage, about 6,000 tons. Her engines are of the inverted three-cylinder type, the high pressure cylinder being 63-ins. diameter, the two low pressure cylinders being 91-ins. each, while the piston stroke is 66-ins. Six double ended boilers and one single ended, having in all 39 furnaces, are fitted. The power expected to be developed is about 9,000 indicated. The speed guaranteed by the builders of the *America* is 18 knots an hour, and confidence is entertained by all concerned as to this result being attained.*

It is abundantly evident, notwithstanding what has already been achieved, that the brisk competition among transatlantic companies for the "fastest steamer afloat" has not yet exhausted itself. The determination some time ago publicly expressed by Mr John Burns, the able chairman of the Cunard Company, to maintain a leading position, has since taken decidedly active shape in the contract entered into and now being carried out by Messrs John Elder & Co. : that is, the construction of the two huge and powerful steamers of unprecedented speed, already referred to near the beginning of this work. They are each of 8000 tons burthen, 500 feet in length, 57 feet broad, by 40 feet depth of hold. Engines of 13,000 horse power

* While these sheets were passing through the press, the *America* was tried inofficially on the Clyde, and attained a speed of 17 knots, with about 6,500 indicated horse-power. On her passage from the Clyde to the Mersey she maintained, it is stated, 18¼ knots over the whole distance.

will be provided, which, it is computed, will drive the vessels at a speed of 19 knots an hour. With the establishment of these remarkable steamships in this most important service, the prospect is near of a transatlantic passage lasting only six days, if not indeed considerably under that period.

Communication with our South African colonies is another service in which modern progress, as regards high speed, has been conspicuously manifest. The steamers engaged in this service—belonging to the Union Steamship Coy. and Messrs Donald Currie & Co.—had special attention directed towards their powers as to fast steaming were exerted to the utmost them during the Zulu War of 1879, at which juncture in the transport of our soldiery. In the autumn of 1878 the *Pretoria*, belonging to the Union Coy., made the outward passage to the Cape, *via* Maderia, in 18 days, 16 hours, including $4\frac{1}{2}$ hours detention. The passage home was made in the autumn of 1879 by the same vessel in 18 days, $13\frac{1}{4}$ hours, including about $5\frac{3}{4}$ hours stoppages. These passages are fairly representative of the best performances of the vessels engaged in this service, and they have not since been much excelled. In midsummer, 1880, the *Durban*, another of the Union Line vessels, accomplished the homeward run *via* Maderia in 18 days, 9 hours, including about $6\frac{1}{2}$ hours stoppages. The *Drummond Castle*, belonging to Messrs Donald Currie & Co.'s Castle Packet line, has made the homeward run in 18 days, 18 hours, or, excluding detentions, in 18 days, 13 hours. The *Hawarden Castle*, of the same line, has made the fastest outward run on record. In the autumn of 1883 she accomplished it in 18 days, 15 hours, including five hours detention at Maderia, leaving the actual steaming time 18 days, 10 hours. The distance traversed by vessels on this service is some 6,000 miles, and the average speed attained is about 13 knots per hour. In the case of one of the Union Coy.'s vessels, the average speed attained has been as high as 13·8 knots per hour over the greater portion of the voyage, the indicated horse-power developed being about 2,570, and the consumpt of coal about $52\frac{1}{2}$ tons per day. For a considerable time recently the Companies have found it more remunerative to drive their vessels at moderate speed, but in times of emergency, such as the outbreak of hostilities in our colonies, their qualities as transports traversing long distances at high speed are eminently efficient.

The employment of steamships in long voyages and at high rates of speed, for which, not so long ago, it was generally supposed sailing ships were only adapted, has been eminently successful. By the opening of the Suez Canal the passage to China was shortened from about 13,500 miles to about 9800 miles, that to India from over 10,000 miles to 6000. Although steamers were running to China *via* the Cape of Good Hope, before the opening of the Canal, and doing the service most admirably, it is subsequent to that great change, and indeed quite recently that the most noteworthy advances have been made in shortening the time occupied on these important services. The passage is now made by steamers under ordinary circumstances in less than thirty days, which sailing ships under the most favourable conditions took three and a half to four months to accomplish. The average speed attained by the steamers prior to the short route never exceeded tén knots; steamers now frequently average twelve knots over the whole distance, except during their passage through the Canal.

The *Stirling Castle*, built in 1882 by Messrs Elder & Co., for Messrs Skinner & Co.'s China fleet, attained a speed of 18·4 knots on her official trial. During 1883 she proved herself to be the fleetest vessel ever engaged in the China tea-carrying trade, arriving in the Thames several days ahead of the China mails, although the latter came part of the way overland. The run from Woosung to London was made in 27 days 4 hours steaming time. Other vessels belonging to this Company, and vessels of the other lines on this important service, although not equalling the performances of the *Stirling Castle*, are exemplifying almost daily the immense superiority of steamers over sailing ships for regularity and despatch in long passages.

As the distance to Australia—*i.e.*, some 12,000 miles as ordinarily taken—is only about 900 miles less *via* the Suez Canal than by the Cape of Good Hope, steamers are employed on both routes. On the 12th May, 1875, the *St. Osyth* left Plymouth for Melbourne *via* the Cape, called at St. Vincent for coal, and thence steamed continuously to Melbourne, reaching her destination on the 27th June. Her full steaming time was about 43½ days, the average speed attained being over 11½ knots per hour. This passage, although considered most remarkable at the time, has since been surpassed. The *Lusitania*, of the Orient line, in 1877 made the passage to Melbourne in 40¼ days, including a detention of 1¼ days at St. Vincent while coaling. Her actual steaming time was almost exactly 39 days, her average speed being only a trifle under 13 knots. The *Cuzco*, of the same line, during the summer of 1879, made the homeward passage from Adelaide to Plymouth in 37 days 11 hours, including all detentions. In the *Orient*, which was the first vessel specially designed and constructed for the Australian direct steam service, a most noteworthy step in advance was made. She was launched in September, 1879, from the yard of Messrs Elder & Co., and on her completion was tried for speed, when she attained a maximum average speed of 17 knots per hour. She has made the passage from Plymouth to Adelaide, *via* Suez Canal, in 35 days 16 hours, and the same voyage *via* Cape of Good Hope in 34 days, 1 hour, steaming time.

The *Orient* was followed in 1882 by the magnificent *Austral*, whose high promise was suddenly blighted for a time by an unfortunate accident. While coaling at her moorings in Sydney harbour by night, the water was allowed to flow into the ship through her after coal ports, carelessly left open and unwatched, and she thus gradually filled, and sank to the bottom. She has since been raised, brought home, and restored to her pristine splendour. She is presently engaged in the express service of the Anchor Line between Liverpool and New York, her performances being such as should gratify all concerned. The *Austral* on her trial attained a speed of 17·3 knots, and has

S. S. AUSTRAL.—ANCHOR LINE.

LENGTH,	...	..	455 ft. 0 in.	DEPTH,	... ...	37 ft. 0 in.
BREADTH,	...	...	48 ft. 0 in.	TONNAGE (GROSS),...		5,588 tons.

BUILT BY MESSRS. ELDER & CO., 1881.

PORTRAIT
AND
BIOGRAPHICAL NOTE.

JAMES ANDERSON.

JAMES ANDERSON, F.R.G.S,

CHAIRMAN OF THE ORIENT STEAM NAVIGATION COY., LIMITED ;
CHAIRMAN OF THE LONDON BOARD OF DIRECTORS OF THE SCOTTISH
PROVINCIAL INSURANCE COY. ; DIRECTOR OF THE HOME AND
COLONIAL INSURANCE COY., DIRECTOR OF THE BANK
OF BRITISH COLUMBIA, ETC.

BORN at Peterhead, Aberdeenshire, on 17th May, 1811, his family then being—and having been since 1780—extensively engaged in shipowning and shipbuilding there. Removed to London in 1831, and entered the counting-house of Mr James Thomson, a considerable shipowner, whose vessels were principally engaged in the West Indian trade. Assumed partnership with Mr Thomson in 1847, carrying on business as James Thomson & Co., a connection which, unfortunately, was soon thereafter broken, in the removal by death of Mr Thomson. In 1849 the business was extended to the Australian trade, by the commencement of a line of sailing vessels to Adelaide, which soon became well-known and favourite traders. Some time after Mr Thomson's death, the name of the firm was changed to Anderson, Thomson & Co., and in 1869 it underwent a second change to Anderson, Anderson & Co., its present designation. In 1876 the feasibility of running a direct line of steamships to Australia occurred to Mr ANDERSON and his partners, and was practically tested at their sole risk in that year. Notwithstanding the predictions that severe loss would result, the experiments encouraged Messrs Anderson, Anderson & Co. to promote the formation of a company to work such a service. Early in 1877, Messrs F. Green & Co. joined Messrs Anderson, Anderson & Co. in the enterprize, and on the 7th March, 1878, the steamer *Garonne* left England for Australia, flying the flag of the ORIENT STEAM NAVIGATION Co., LIMITED, the designation "Orient" having been adopted through the high reputation of the clipper ship of that name belonging to Messrs Anderson, Anderson & Co. Anticipations were at first confined to the hope that sufficient trade might be found to justify monthly sailings, but almost at once it was seen that a fortnightly service was requisite. At the outset four steamers—the *Chimborazo, Lusitania, Cuzco,* and *Garonne*—were purchased by the Company, and one—the *Orient*—built. In January, 1880, the Pacific Steam Navigation Company entered, as it were, into partnership, by supplying, in ready and admirable working order, the additional vessels required. The further additions to the fleet, and the nature of the service done, are referred to elsewhere in this work.

James Anderson

made the passage from Plymouth to Melbourne, *via* the Suez Canal, in the unprecedented time of 32 days, 14 hours steaming.

Until quite recently the only direct communication with New Zealand has been by sailing vessels, but the New Zealand Shipping Company (Limited) and the Shaw, Savill, & Albion Company (Limited) are at the present moment in the thick of organising monthly services of high-class modern steamships to the Antipodes. The former Company in 1883 despatched the *Ionic*, which they had chartered, with other of the White Star steamships, for the purpose. This vessel made the passage out to New Zealand in 43 days, and home in 45 days, including stoppage for coaling. Passages of a similar character have been made by this vessel and others of the Company's own fleet, three of which—the *Tongariro, Aorangi*, and *Ruapehu*— are splendid new steel vessels from the stocks of the famous Fairfield yard. The vessel last named has just made the passage home from Lyttelton, New Zealand, to Plymouth, in the marvellously short period of 37 days, 20 hours, 40 minutes, steaming time; the time, with detentions, being about 39 days. The other Company referred to are having two magnificent steel vessels built by Messrs Denny & Bros., of Dumbarton, to be named the *Arawa* and *Tainui*, each of 5000 tons gross. These vessels are to maintain a sea speed of 12½ knots, the engines to be fitted representing a noteworthy advance in the line of economical consumpt of fuel with prolonged terms of steaming.

Between 1875 and 1882 the number of steamers having ocean speeds of 13 knots and upwards, increased from twenty-five to sixty-five. Of these there were only ten—previous to 1875—of 14 knots speed and upwards, whereas at the beginning of 1882 there were twenty-five of this character. During the years 1882 and 1883 alone the increase in the number of such vessels has been almost double that for the previous period named. The highest speed previous to 1875 did not exceed 15 knots, now there are numerous vessels with speeds exceeding 17 knots, several even approaching 18 knots, while

in one or two cases the speed attained—under favourable
circumstances probably—is stated to have been considerably
over 18 knots, the Guion Liner *Oregon,* indeed, reaching the
round figure of 20 knots.

Viewed purely from the point of view of the sea voyager,
such results are alike remarkable and gratifying, whilst considered in their technical and commercial aspects they also
call for admiration. It is questioned, however, whether in
most cases the attainment of great speed has been accompanied
with corresponding or proportionate advance in other matters
with which vital progress is concerned. Commercially, it is
of the utmost importance that increase of speed and power
should be achieved, with the least possible weight of machinery,
water, and fuel to be carried; with the least possible expenditure of fuel; with safety and efficiency in working; with low
wear and tear, and cheapness of maintenance.

The efficiency of the ship and machinery in fulfilling the
various and often conflicting conditions of economical service is
a matter with which the naval architect and the marine engineer
have jointly to deal. Where the conditions cannot all be equally
satisfied, it is the province of these two to make that sort of
compromise which gives the best results in each special case.
In cargo-carrying vessels, for example, an economy in the
consumption of fuel may often be the dominant and regulating
quality. An economy of one-fourth of a pound per horse-
power per hour gives, on a large transatlantic steamer, a saving
of about 100 tons of coal for a single voyage. To this saving
of cost is to be added the gain in wages and sustenance of
the labour required to handle that coal, and the gain by 100
tons of freight carried in place of the coal. Again, it is
estimated that every ton of dead-weight capacity is worth on
an average £10 per annum as earning freight. Supposing,
therefore, the weight of machinery and water in any ordinary
vessel to be 300 tons, and that by careful design and judicious
use of materials the engineer can reduce it by 100 tons without
increasing the cost of working, he makes the vessel worth
£1,000 per annum more to her owners. To these and other

such considerations, which often influence the naval architect and engineer in their designs, and due regard to one or more of which not infrequently prevents the attainment of all-round success, should be added many others concerned with the after-management of vessels. For example, the length of voyage to be performed, the seasons and the markets in particular trades, the number of ports of call, and the coaling facilities at each, are all matters which must be taken into consideration when measuring, from one standpoint or from particular instances, the degree of success attained in general.

The diminution in coal consumpt, coincident with the increase of steam pressure and the acceleration in speed which has been attained in recent years measures the principal element of progress. In many of the "racers" of recent times, it is true, speed is attained at what may appear a great sacrifice of fuel, but these are cases in which the commercial considerations often used to measure the efficiency of ordinary cargo-carrying steamers are not applicable. Owners—of transatlantic steamships especially—realise from experience that "speed pays," and they find it of more advantage to ensure certainty of arrival at the port of destination than to save a few tons of coals on the voyage.

During the past sixteen years or so the advance made in respect to the reduced ratio of fuel consumed to power developed has indeed been considerable. Before the period stated a vessel of say 700 tons carrying capability was not only much slower than the present-day vessels but the coal supply amounted to about 16 tons per day of 24 hours, whereas vessels are now being built of like size which attain an average speed of 9 knots, the consumpt of coal not being more than 6 tons per day. In 1872 the consumption of coal in vessels whose engines were worked at a pressure of from 45-lbs. to 65-lbs. per inch (the latter being then the highest pressure recorded), did not exceed 2½-lbs. per indicated horse power per hour. This indicated an improvement in the marine engine during the previous decade, represented by a reduction in the consumpt of fuel by more than one half the amount previously thought indispensable.   Since 1872, there has been a further reduction in the average consumpt of fuel to the extent of 15 or 16 per cent., or in the average from $2\frac{1}{8}$-lbs. to less than $1\frac{3}{4}$-lbs. per indicated horse power per hour.

As in the case of the vessels themselves, mild steel is largely taking the place of iron in the construction of marine boilers. The change has reduced the weight of this important item of machinery by about one-tenth, a great advantage in itself, as increasing the dead-weight capability of the vessel.   The questions as to the reliable character of the boilers made of steel with respect to strength under working, and as regards corrosion, are being practically answered as time goes on; and, as in the case of ship structure, in a way very satisfactory for the new material.   There is every probability that a further advance may soon be made in connection with marine boilers, in the way of constructing the shell in solid rings, thus doing away with the longitudinal seams.   The strength of boilers is of course governed by the strength of the seam, and this is never above 75 per cent. of the solid plate.   Hence, if solid shells are employed, an increase in pressure of about 25 per cent., with the same thickness of shell, may be obtained. Appliances are now being laid down in the Vulcan Steel and Forge Coy., Barrow-in-Furness, for this purpose.

Improved appliances and modes of construction, no less than the change of material employed, have played a large part in rendering the boilers of modern steamships capable of being worked at the higher pressures now common.   It is not possible, however, with the space at command, to treat of these; nor is it practicable to consider or even enumerate all the various improved fittings which in the aggregate so materially enhance the efficiency of boilers.

One such feature particularly noteworthy because of the success with which it has been applied to the boilers of very many modern high-class merchant ships may be shortly referred to.   This is the corrugated mild steel furnace, manufactured by the Leeds Forge Company on Mr Samson Fox's patent, an illustration of which is given in Fig. 4.

This shows a single corrugated furnace flue, flanged at the end to meet the tube plate of the boiler. The strength of these flues to resist collapse has been proved in the presence of the officials of the Admiralty, Board of Trade, and Lloyd's Register, to be, on the average, four times greater than a plain flue of the same dimensions. An immediate effect of this has been to increase their average diameter from 3-ft. to 4-ft., the thickness of plate—½-inch—remaining the same; a result as to diameter and thickness quite impracticable with ordinary furnaces. Some have even been made to carry 170-lbs. per square inch of steam pressure, 4-ft. 8-ins. outside diameter

FIG. 4.

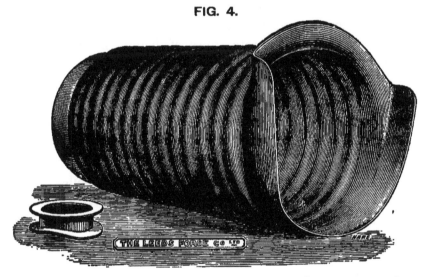

THE LEEDS FORGE CO LD

constructed of one single plate, with the weld so arranged as to be below the fire bars in the furnace.

By the corrugated, as against the plain tube, a greatly increased heating surface is presented to the flame and the heated gases of the furnace, thus yielding a greatly enhanced evaporative power, equal to at least 50 per cent. more than in the ordinary form. Better allowance is made by the corrugated surface for the expansion and contraction caused by changes of temperature in the furnace, without in any way impairing its efficiency as a longitudinal stay for the boiler. Through the increased diameters and the augmented surface possible by these corrugated tubes, their adoption lessens the number of furnaces and stokers necessary for the horse-power required. As a further consequence, the boiler space may be diminished, and an increase effected in the cargo space or freight-carrying capacity of the vessel.

The advantages of corrugated flues as compared with plain flues cannot all be named, but the extraordinary extent to which they are now employed in the best class of steamships is the best proof of their superiority. It is stated that if the flues which have been made by the Company since their introduction about the beginning of 1878, and are now at work, were placed in one continuous line, they would extend to a length of over twenty miles, representing, in marine and other engines, nearly one million horse-power.

The number of separate types of boilers introduced into steamships has been much increased of recent years—an evidence that engineers are growingly conscious of the possibilities which may result from improved efficiency in this agent of propulsion. One direction in which their efforts at present are being largely put forth, is that of securing the more complete combustion of fuel in the furnaces. Considerable success has already attended the working of boilers under forced draught, or the admission of air to the furnaces under pressure. Combined with special types of boilers, it has been affirmed that nearly 50 per cent. more power has been obtained by this means. There is doubtless much to be expected from this system in the future, especially as it may be associated with a change in the form or type of boilers by which the number and weight of such items will be reduced. The saving of space in the vessel, the economy in comsumption of coal, the reduction in dead-weight of machinery, are possibilities of the movement now in progress which cannot fail to effect materially the commercial character of our high-class mail and passenger steamships, and merchant vessels generally.

Other directions in which advance has been made during the period under review are, considerably higher steam pressures, less heating surface, and smaller cylinders, for indicated horse-power developed. The various improvements in design and construction which have contributed to these results cannot be entered into with any degree of fulness here. For detailed treatment of these matters, readers are referred to the papers read by eminent engineering authorities, before the various professional and scientific institutions, a list of which papers follows the present chapter.*

Reduction in the weight of machinery per indicated horse-power developed is, in general terms, the common line in which engineering effort lies, and in which no little advance has lately been made. Every possible opportunity of using steel, where it can be introduced with safety and efficiency, has been taken advantage of. Hollow crank steel shafts and pro-peller shafting in place of solid shafting; propellers and pistons of cast-steel in place of iron; and boilers of mild steel plates, are a few of the directions in which large weight-savings have been effected. That there is still great room for improvement in this direction is shown by the following statement, given by Mr. F. C. Marshall, of Messrs R. & W. Hawthorn, Newcastle-on-Tyne, in his valuable paper read before the Institution of Mechanical Engineers in 1883. The figures given show for various classes of vessels the average weight of machinery per indicated horse-power, in steamships of the merchant marine —and for comparison—of the Royal Navy:—

	Lbs. per I.H.P.
Merchant Steamers, - - - - -	480
Royal Navy, - - -	360
Royal Navy, fast cruiser *Iris,*	280
Torpedo Ram, *Polyphemus,* - -	180
Torpedo vessels, - - -	60
Ordinary marine boilers, including water, - -	196
Locomotive boilers, including water, - -	60

The figures given are for weights of machinery, including engines, boilers, water, and all fittings ready for sea.

---

* This list with those which follow other chapters, have been compiled at considerable trouble in the hope that they may be of use to technical readers in directing them at once to accurate and detailed information. In this con-nection also, the excellent work by Mr. A. S. Seaton, "Manual of Marine Engineering," and that by Mr. W. H. White, "Manual of Naval Architecture," may be referred to with every satisfaction.

One of the most important of recent advances in marine engineering—affording as it does the means of using higher steam pressures than have hitherto been used with economy —is the introduction of the triple expansion description of engines already referred to. This important departure was begun in 1874, when Mr A. C. Kirk, of Messrs R. Napier & Sons, designed and fitted on board the screw-steamer *Propontis*, built for Mr W. H. Dixon, of Liverpool, by Messrs Elder & Co.—with whom Mr Kirk at that time was engineering manager—engines involving the principle of triple expansion and abnormally high pressure of steam. In 1877 the principle received further practical development on board the *Isa*, a pleasure yacht fitted with triple expansion engines, designed in 1876 by Mr Alexander Taylor, consulting engineer of Newcastle-on-Tyne, who has subsequently designed several other engines of the same type for larger merchant steamers.

As not infrequently happens in connection with inventions, several minds were occupied, and independent ideas matured almost simultaneously, in the matter of triple expansion engines. Mr Kirk had secured the patent for engines involving this principle subsequent to, but before he was made cognisant of, Mr Taylor's work. At the same time he learned that in quite another quarter the designs for such a type of engine had already been perfected. Mr Kirk, on hearing these facts, relinquished the patent rights he had secured. Notwithstanding this, it is to the success of the engines designed by Mr Kirk, and fitted by his firm on board the screw-steamer *Aberdeen*, that the recent development of the system is largely due. This vessel was built in 1881 for the Australian service of Messrs G. Thomson & Co., London and Aberdeen, and measures 350 feet by 44 feet by 33 feet. Her engines work at a boiler pressure of 125 lbs. per square inch. The three cylinders are respectively 30 inches, 45 inches, and 70 inches in diameter, and the stroke is 4 feet 6 inches. The smallest is the high pressure cylinder, into which the steam is first admitted; from thence it passes, after expansion, into the second or intermediate cylinder; after still further expansion

PORTRAIT
AND
BIOGRAPHICAL NOTE.

ALEXANDER C. KIRK.

it passes into the third or low pressure cylinder, from whence, after the expansion is completed, it is discharged into the condenser.

When the *Aberdeen* was completed, 2,000 tons of dead-weight were put on board, and the consumption was tested on a four hours' run at 1,800 horse power. The result was the consumption at the rate of 1.28-lbs. per indicated horse-power per hour, with Penrikyber Welsh coal. From this the designer of the engine inferred a sea consumption of good Welsh coal at the rate of 1·5 to 1·6-lbs. per indicated horse-power. The maximum measured mile speed was 13·74 knots, with 2,631 indicated horse-power, and a consumption of 1 ton 17 cwt. per hour. The vessel started from Plymouth on 1st April, 1881, upon her first voyage to Melbourne, with 4000 tons of coals and cargo—weight and measurement—on board. She arrived at Cape Town on the 23rd April, having accomplished the distance—5,890 miles—in 22 days. After taking in about 140 tons of coal, she left for Melbourne on the 24th, and arrived there on the 14th May, in 20 days. The whole time occupied in steaming from Plymouth to Melbourne was, therefore, 42 days. Her average indicated horse-power on the voyage has been about 1,880, and the consumption less than thirty-four tons per day, or at the rate of about 1·69-lbs. per indicated horse-power over the whole voyage. Since these results were obtained, Messrs Napier have fitted three sets of 5000 H.P. triple expansion engines into vessels built for the Compania Transatlantica Mexicana, and are completing a duplicate of the *Aberdeen*.

The firm of Messrs Denny & Coy., Dumbarton, are at present making engines of the triple expansion type for the new steamers of the Shaw, Savill & Albion Company's direct New Zealand service. There are four cylinders and two cranks, the cylinders being arranged in pairs, tandem fashion, the small on the top of the large. Expansion takes place in three stages, the first small cylinder taking steam from the boilers about five-eights of the stroke, and expanding into the valve chest of the second small cylinder, where it is further expanded.

From thence it exhausts into the valve chest common to both
the large cylinders described.   The steam to be supplied to
these engines is to have a pressure of 160-lbs. per square inch,
the highest yet carried in marine engines.   These instances
of actual advancement, taken in conjunction with the favourable light in which the triple expansion principle is regarded
by our foremost marine engineers, augur well for the future
of steamship propulsion.

The activity characterising merchant ship construction,
and especially the enormous increase in their dimensions
and speed within recent years, have necessarily led to specu-
lation with regard to what form the ship of the future will
take.   There have not been wanting, indeed, actual propositions
and elaborately prepared designs of what the ideal ship should
be.   A company was sometime ago formed in Washington,
U.S., to have three vessels built of a novel type, the patented
invention of Captain Lundborg, a Swedish engineer, intended
to make the Atlantic passage in five days.   It was also
announced that the order for their construction had actually
been given out, but this is wanting in confirmation.   Great
expectations were entertained in America regarding what was
termed the dome-ship *Meteor*, built on the Hudson in the
early part of 1883 from the designs of Captain Bleven.   A
company had been formed under the designation of the
" American Quick Transit Company," the chief supporters
being Boston merchants, to build several large steamships on
the proposed lines, but the utter failure of the *Meteor* to answer
the promises of her inventor has relegated the scheme to the
vast limbo of unfulfilled American projects.   Three years ago
or more, scientific journals gave publicity to a scheme of "Ocean
Palace" steamship, patented by Mr Robert Wilcox, of Melbourne, Victoria, the claims for which ranked themselves under
the heads of speed, safety, and comfort.   Double hulls, as in
the case of some Channel steamers, were employed, but each
of the hulls was divided into two cigar-shaped portions, thus

giving to the submerged whole, a quadruplicate character, and which, with its palatial superstructure, was apt to remind one —shall it be said?—of Rome and her seven hills, or Venice and her island base! The design, nevertheless, was to give the least resistance with the greatest buoyancy and stability The method of propulsion proposed by Mr Wilcox was also novel. He placed a couple of enormous drums fore and aft (between the hulls), which were to be driven by the engines as if they were paddle-wheels. Over these drums was placed a continuous band of iron links, upon which, at suitable intervals, paddles or blades were fixed. A comparatively low speed of engine was to give a high speed of velocity to this band of blades; and as there would be twenty-one paddles, all immersed at the same time, their grip of the water was to be such that there should be little slip. Whether on a serious application of the principles involved in this invention to a ship for the Australian service the voyage would have been made, as was claimed, in 26 days, equal to an increase in speed of 75 per cent., has never of course been determined! Still another scheme, and one which the inventor has been encouraged to prosecute by the recommendations of eminent authorities on both sides of the Atlantic, is that of Captain Coppin, noted for his success in salvage operations, which consists of an "Ocean Ferry" partaking as to form somewhat of the features above described for Mr Wilcox's "Ocean Palace." The speed said to be possible by Captain Coppin's vessel is twenty knots an hour, and the terminal ports proposed are Milford Haven and New York. It was announced some time ago that M. Raoul Pictet, the eminent engineer of Geneva, was engaged upon the question of ship design and propulsion, and was in hopes that by application of his ideas he might yet send ships careering over the sea at the rate of thirty-seven miles an hour!

Enough has been said to show that there is no lack of inventive effort being put forth towards a realization of the ideal ships of the future. In a service, however, like that of the Atlantic, where competition is strong and keen, and where the monetary issues are neatly adjusted between rival companies, there is little chance of any of the various projects being tried. An impression exists among shipowners—for which doubtless there are sufficient grounds—that time and capital staked on novelties or "new departures" are simply invitations to defeat in the race or to absolute ruin itself. This commercial prudence and industrial caution has been startled in several ways of recent years—*e.g.*, by meteoric flashes such as the *Livadia* and *Meteor*—the ultimate effect of which has been to illumine and make clearer the probable line of advancement.

By pretty general consent of those most competent to judge the ships of the immediate future will possess the broad distinctions of being either purely passenger or purely cargo-carrying mediums. It is equally agreed that twin in place of single screw propellers will be employed, and that for the express ships nothing less than 20 knots per hour will be considered satisfactory. On a subject, however, concerned not with historical facts, but with theories and scientific forecasts, it may be well not to enlarge, especially as the future is evidently charged with possibilities of which present-day designers can have but indefinite notions. The subject of employing electrical energy as the propulsive power on board ship is at the present time engaging serious attention, but the degree of practical and commercial success attained does not, as yet, warrant any anticipation of its immediate application to vessels beyond small craft, such as launches and ferries. In the midst, however, of such immense and marvellous works achieved by this great—and, in some senses, modern—force, it would be both idle and unwise to keep out of view the possibilities of its future as affecting ship propulsion.

List of Papers and Lectures bearing on the speed and propulsive power of modern steamships, to which readers desiring fuller acquaintance with the *technique* and details of the subject are referred :—

ON THE BOILERS AND ENGINES OF OUR FUTURE FLEET, by Mr J. Scott Russell: Trans. Inst. N.A., vol. xviii., 1877.

On the Compound Marine Steam Engine, by Mr Arthur Rigg : Trans. Inst. N.A., vol. xi., 1870.

On Compound Engines, by Mr Richard Sennett : Trans. Inst. N.A., vol. xvi., 1875.

On the Progress Effected in the Economy of Fuel in Steam Navigation, Considered in Relation to Compound Cylinder Engines and High Pressure Steam, by Mr F. J. Bramwell ; Proceedings Inst. Mech. Engineers, 1872.

Our Commercial Marine Steam Fleet in 1877, by Mr J. R. Ravenhill : Trans. Inst. N.A., vol. xviii., 1877.

On the Steam Trials of H.M.S. *Iris*, by Mr J. Wright : Trans. Inst. N.A., vol. xx., 1879.

On the Steam Trials of the *Satallite* and *Conquerer* under Forced Draught, by Mr R. J. Butler : Trans. Inst. N.A., vol. xxiv., 1883.

On Combustion of Fuel in Furnaces of Steam Boilers by Natural Draught, and by Supply of Air under Pressure, by Mr James Howden : Trans. Inst. N.A., vol. xxv., 1884.

Propositions on the Motion of Steam Vessels, by Mr Robert Mansell : Trans. Inst. Engineers and Shipbuilders, vol. xix., 1875-76.

On Steamship Efficiency, by Mr Robert Mánsel : Trans. Inst. Engineers and Shipbuilders, vol. xxii., 1878-79.

The Comparative Commercial Efficiency of some Steamships, by Mr Jas. Hamilton Jun.: Trans. Inst. Engineers and Shipbuilders, vol. xxv., 1881-82.

The Speed and Form of Steamships Considered in Relation to Length of Voyage, by Mr James Hamilton, Jun. : Trans. Inst. N.A., vol. xxiv., 1882.

On the Comparative Efficiency of Single and Twin Screw Propellers in Deep Draught Ships, by Mr W. H. White : Trans. Inst. N.A., vol. xix., 1878.

On Twin Ship Propulsion by Mr G. C. Mackrow : Trans. Inst. N.A., vol. xx., 1879.

On Marine Steam Boilers : their Design, Construction. Operation, and Wear, by Mr Charles H. Haswell : Trans. Inst. N.A., vol. xviii., 1877.

On the Introduction of the Compound Engine and the Economical Advantages of High Pressure Steam, by Mr Fred. J. Rowan : Tran. Inst. Engineers and Shipbuilders, vol. xxiii., 1879-80.

On Compound Marine Engines with Three Cylnders Working on Two Cranks, by Mr Robert Douglas : Trans. Inst. Engineers and Shipbuilders, vol. xxv., 1881-82.

On the Triple Expansive Engines of the s.s. *Aberdeen*, by Mr A. C. Kirk : Trans. Inst. N.A., vol. xxiii., 1882.

On the Efficiency of Compound Engines, by Mr W. Parker : Trans. Inst. N.A., vol. xxiii., 1882.

On the Construction and Efficiency of Marine Boilers, by Mr Josiah M'Gregor : Trans. Inst. Engineers and Shipbuilders, vol. xxiii., 1879-80.

On the Strength of Boilers, by Mr J. Milton : Trans. Inst. N.A., vol. xviii., 1877.

On the Use of Steel for Marine Boilers and some Recent Improvements in their Construction, by Mr W. Parker: Trans. Inst. N.A., vol. xix,, 1878.

ON THE REACTION OF THE SCREW PROPELLER, by Mr James Howden : Trans. Inst. Engineers and Shipbuilders, vol. xxii., 1878-79.

ON THE PROGRESS AND DEVELOPMENT OF THE MARINE ENGINE, by Mr F. C. Marshall. Proceedings Inst. Mech. Engineers, 1881.

ON SOME RESULTS OF RECENT IMPROVEMENTS IN NAVAL ARCHITECTURE AND MARINE ENGINEERING, by Mr William Pearce. Lectures on Naval Architecture and Marine Engineering : Glasgow, William Collins & Sons, 1881.

THE SPEED AND CARRYING OF SCREW STEAMERS, by Mr William Denny. Lecture delivered to the Greenock Philosophical Society, 20th January, 1882, in honour of the birthday of James Watt (19th Jan.) : Greenock, Wm. Hutchison.

ON THE ADVANTAGES OF INCREASED PROPORTION OF BEAM TO LENGTH IN STEAMSHIPS, by Mr J. H. Biles : Trans. Inst., N.A., vol. xxiv., 1883.

CAST STEEL AS A MATERIAL FOR CRANK SHAFTS, by Mr J. F. Hall, Inst. N.A., vol. xxv., 1884.

CHAPTER III

EVERY advance—whether it be in dimensions or power of steamships, or whether it consist of modifications in their structure or appointment—toward that ideal period when sea-voyaging will have attained its maximum of comfort and its minimum of risk, is deserving of record. The qualities of safety and comfort, even more than increase of speed and the consequent shortening of sea passages, are first essentials in the realisation of this great end. The structural modifications, and the great development in size of recent vessels, affect the qualities named in ways which already may have been made evident, but which call for more detailed treatment. The more minute watertight sub-division of the hulls of vessels, for instance, and especially the presence of an inner skin or cellular bottom, are marked accesions to their safety.

The primary object and ruling principle of all proper watertight sub-division, is so to limit the space to which water can find access, that in a vessel with one, or even two, compartments open to the sea, the accession of weight due to the filling of these compartments would not exceed the surplus buoyancy she should possess. Until within recent years this was not so fully regarded as it ought, owing chiefly to the objections of shipowners to minute sub-division, as impairing a vessel's usefulness and capacity for stowage of miscellaneous cargo. These objections have still doubtless much weight for vessels in certain trades, but the tendency of modern passenger traffic to estrange itself from cargo-carrying mediums, makes them almost inapplicable to a large section of our mercantile marine. There is now, indeed, more

faith in well divided ships generally as being in the long run no
less efficient and more economical than scantily divided ones.

The salutary influence exerted by the Admiralty, in stipu-
lating for increased sub-division of the hulls of all merchant
vessels eligible for state employment in times of war, is

FIG. 5.

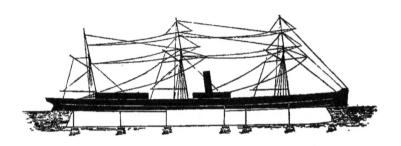

worthy of special recognition.   A few years ago only
thirty or forty large steamers in the merchant navy were so
constructed, as regards sub-division, that they would have
survived for a few minutes the effect of collision with other
vessels or of grounding on rocks.   Within recent years—
greatly owing to the stipulations referred to, and to the desire

FIG. 6.

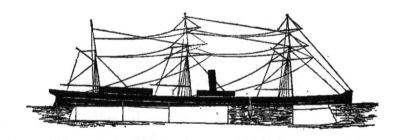

of shipowners to comply with them for the reasons given—
there are few, if any, of the many first-class mail steamers
turned out, not so constructed.

Much valuable information on the subject was given in a
paper on "Bulkheads," read before the Institution of Naval
Architects in March, 1883, by Mr James Dunn, of the
Admiralty, whose experience in matters relating to the quali-

fication of merchant ships for State employment eminently
entitles him to be considered an authority. From diagrams
contained in the paper, the effects of good and of inefficient
sub-division of vessels are well illustrated. Figs. 5 to 8 in
the present work represent some of these. They are concerned

FIG. 7.

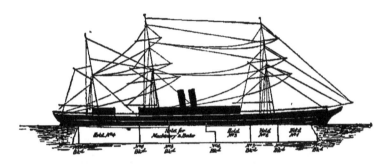

with two vessels, in one of which—an actual case—the bulkheads were well placed and cared for, and carried to a reasonable height as shown in Fig. 5; the result of a collision
proving that under such conditions they were of immeasurable
value, while in the other vessel, although having the same
number and a similar disposition of bulkheads, their presence

FIG. 8.

is rendered valueless by their being stopped at or about the
water-line, as indicated in Fig. 7. In the first case, a steamer
of nearly 5,000 tons, during a fog, ran into the vessel represented by Fig. 5 and 6, striking her abreast of No. 3 bulkhead,
and opening up two compartments to the sea. The bulkheads,
however, as has been said, were carried to a reasonable height,

and the water could not get beyond them—they stood the test—the vessel did not sink, but kept afloat at the trim shown in Fig. 6, and in this condition steamed 300 miles safely into port. The second case—though a suppositionary one merely, yet representative of not a few merchant steamers now afloat—would not be attended with like results should such an accident happen as has been described. In vessels so bulkheaded, the water not being confined to the two holds, numbered 2 and 3, as it was in the previous actual case, would pour over the top of the dwarf bulkhead into the foremost hold, and the ship would soon assume the position indicated in Fig. 8: one not at all favourable, as may be readily believed, for the completion of a voyage to port.

These cases illustrate the value of minute and careful sub-division of the hulls of vessels by watertight bulkheads. Unless, however, the bulkheads are carried a few feet higher than the level of the water outside—and it is to be regretted that this is still not infrequently overlooked or neglected in merchant steamers—they are valueless, and, indeed, had better not be in the ship at all. They will contribute to the loss of the vessel by keeping the water at one end, and carrying her bows under, whereas if they are not fitted, the same volume of water will distribute itself throughout the bottom of the ship fore and aft, preserve the even trim of the vessel, and allow more pumps to cope with the inflow. Although her freeboard, or height of side above water will be reduced, she will still be seaworthy, the boiler fires may be kept burning, and the machinery going, sufficiently long for her to reach a port of safety. Readers appreciating the above considerations will readily see why it is that sailing vessels are usually fitted with only one transverse bulkhead—that near the bow—and understand how it is that the outcry sometimes made by inexperienced people about the absence of other bulkheads in emigrant sailing vessels is for most part unheeded by those on whom the responsibility falls.

From statistics presented in the paper above referred to, it is shown that during a period of six years, ending with

December, 1882, the average loss per annum of ships not qualified for the Admiralty list was one in twenty-five; while of ships so qualified the annual average loss was only one in eighty-six. The chances of loss from any cause are thus seen to be nearly four times as great for a ship not constructed to qualify for the Admiralty list as for a vessel entered on that list. During the first four-and-a-half years of the period referred to, not one ship of those entered on the list was lost by collision although a considerable number had been in collision, and escaped foundering by reason of the safety afforded by their bulkheads. During 1882 six casualties happened to ships on the list, one of which—a case of collision —proved fatal. This was a case, however, such as no merchant steamer afloat at the time would have been capable of surviving. The whole of the ship—a small one—was flooded abaft the engine-room, the two after holds being open to the sea. The whole of the losses from the Admiralty list during the period referred to—eleven in number—have been from drifting on rocks, or otherwise getting fixed on shore, with the solitary exception above quoted. In the same period 76 ships have been lost which had been offered for admission to the list, but had not been found qualified; of these 17, or 22½ per cent., were lost by collision; and 10, or 13¼ per cent., were lost by foundering; most of the rest stranded or broke up on rocks. The risk of fatal collision, according to Mr. Dunn, is about 1 to 100, irrespective of the class of ship, and the ships on the Admiralty list enjoy almost absolute immunity from loss by this cause.

The foregoing indicates the way in which minute watertight sub-division has come to be widely regarded. Much requires yet to be done to reach the end desirable, as there are many vessels built prior to the movement sadly deficient in the qualities concerned. The bulkhead near the bow—the "collision" bulkhead, as it is termed—has done noble service in many cases of collision, and it is with reason that its position and structural character in all vessels are subject to special supervision and made a condition of classification in the Registries. Recently it has been made imperative by Lloyd's Society that vessels over 330 feet long should have two additional water-tight bulkheads extending to the upper deck, in the holds, forward and aft of the machinery compartment. The requirements of this Registry, it may be said, constitute at once an anticipation and a reflex of the needs of merchant ship construction. In watertight sub-division, as in other matters, the Society and its large staff of able surveyors are " powers which make for " sterling efficiency.

The extended adoption of double bottoms is specially contributory to the safety of vessels in the event of their running over a reef into deep water, or in going ashore. Numerous instances are on record of steamships so constructed sustaining damage to the outer skin, and yet—because of the inner bottom remaining intact and perfectly water-tight—no serious damage resulting. The case of the *Great Eastern* is an early yet notable example. This great vessel in 1860 ran over a reef of rocks and tore a hole 80 feet long and 10 feet wide in her outer skin, yet, because of this feature in her construction, she was placed in no jeopardy.

In this connection it would seem that even the employment of steel as the constructive material affords safety to a vessel in circumstances which would almost prove fatal to a ship built of iron. The remarkable experience which befell the first steel ocean-going steamer—the *Rotomahana*, belonging to the Union Steamship Company of New Zealand—may here be recounted. While steaming between Auckland and the Great Barrier Island on New-Year's-Day, 1880, this vessel struck upon and ran over a sunken rock. She had a large party of pleasure seekers on board, and but for the fact that she was built of such a ductile material as mild steel, the commencement of the year 1880 might have been clouded by a catastrophe which would have spread gloom and sorrow throughout New Zealand, if not over a wider circle. At the earliest possible moment the damaged vessel was docked for examination. The results are effectively summarised in an extract from a letter referring to the accident, written by the

managing director of the Company.   He says:—"This experience has clearly shown the immense superiority of steel over
iron.   There is no doubt that had the *Rotomahana* been of
iron, such a rent would have been made in her, that she would
have filled in a few minutes."   The starboard bilge for over
20 feet of its length was more or less indented, one plate
especially being greatly misshapen between two frames.   This
plate was removed, hammered, rolled flat again, and replaced
—after the frames which had been bent inwards by the force
of the grounding had been straightened.   No new material
except rivets were required for the execution of the repairs.
The *Rotomahana*, as if to show her ability to "laugh at all
disasters," has grounded twice subsequently on the rocky and
treacherous coast along which she plies, yet has come out of
the ordeal with immunity from positive danger.   Her remarkable experience may safely be taken as most convincing evidence
of the suitability of mild steel for shipbuilding.   Other cases
are not wanting, however, in which the same thing is exemplified.   One which recently astonished everybody concerned
with shipping was that of the *Duke of Westminster*, a vessel
400 feet in length, built of mild steel by the Barrow Shipbuilding Coy., which lay bumping for a week on stony ground
near the Isle of Wight, without making a drop of water.
The bottom plating of the *Duke of Westminster*, as she appeared
in dry dock, was corrugated between the frames for more than
half the length of the vessel, and yet not a single plate was
cracked, nor a rivet started.   Another case of an equally
striking character is that of the British India Coy.'s steamer
India, built by Messsrs Denny, of Dumbarton, which went
ashore near the mouth of the Thames in December, 1881, and
was left high and dry at low water.   Her bottom, although
forced up about 3 inches over a length of about 40 feet amidships, did not give way, and the vessel, during the period she
was aground, did not make a drop of water.

All these are instances of the enhanced safety of ships due
to the employment of steel, which ought certainly to be recognised by underwriters in the way of reduced premiums for

vessels constructed of this material. One consideration which, it is both curious and sad to say, militates against this result, and which, judging from views entertained by shipowners themselves, stands in the way of the employment of steel, is not its *inability* but its very *efficiency* to withstand the results of grounding or other catastrophe. It is argued that while the effects of grounding are less *severe* in the case of steel, and do not result in fracture or through-piercing because of its great ducility, yet the amount of *damage requiring repair* is invariably much greater than in the case of iron. This view of the matter—which virtually places pounds, shillings, and pence before the comfort, if not the very lives, of those on board ship—the author feels bound to say, is not, so far as he knows, shared by owners of ships engaged in mail and passenger service, and it cannot surely be entertained by underwriters of any proper discernment.

Safety in ocean steamships, in so far as affected by design, has unquestionably received greater attention at the hands of designers within recent years than formerly. The particular directions in which this is evinced, as well as the causes at work in bringing it about, will be dealt with in the chapter on scientific progress, the object here being to indicate the extent to which the safety of ships is affected by the qualities of their construction and outfit. The general question of seaworthiness, affected as it is by matters almost beyond the province of the marine architect, is in great measure the care of others concerned. The underwriting or insurance societies looking to their own interests, the Board of Trade on behalf of the lieges, and shipowners on their own and their customers' and servants' account, are parties on whom responsibility devolves in this connection. The question whether they are duly, and at all times alive to such responsibility, is one very difficult to answer, and cannot be fully dealt with here. Apart from the question of remissness by these bodies, in what are clearly their special duties, there is great difficulty in appor. tioning the duties and responsibility aright. The Board of Trade have not infrequently received checks when with precautionary motives they have interfered with departments and in matters but little affecting a vessel's seaworthiness. The conflict which has so long raged and still rages between the Board and the shipowners of Britain regarding the loading of vessels, illustrates, and is indeed the result of, both difficulties. The Merchant Shipping Bill, introduced by Mr. Chamberlain, and in a modified form now before Parliament, will, it is hoped, furnish a satisfactory solution of the matter. Shipowners themselves have too often insisted on exercising functions and dictating in matters which only may be determined with propriety and safety by builders or by competent naval architects.

The amount of thorough supervision to which a vessel is subjected while under construction, renders the fear of unseaworthiness, from either defective construction or equipment, the least reasonable of all the fears with which ocean navigation is regarded. It is in later circumstances, and concerning matters of a more extraneous character, that the most justifiable fears may be entertained regarding a vessel's safety. Overloading, improper stowage, bad management, under-manning, insufficient repair, besides the numerous inevitable and unforeseen circumstances incidental to sea-voyaging, may be instanced as the causes to which the greatest losses are attributable.* Few instances of loss from structural defects are adduceable, and even in these, causes of a more or less extraneous character are associated with the loss. On the other hand, instances could be multiplied where vessels sustaining the casualties

---

* For full and excellent treatment of this subject, see the paper on " Causes of Unseaworthiness in Merchant Steamers," by Mr Benjamin Martell, Chief Surveyor to Lloyd's Register, with the ensuing discussion : Trans. Inst., N.A., vol. xxi., 1880.

Several of the causes above named it is doubtless the province of the scientific shipbuilder, and the duty of the shipowner, to obviate by furnishing the captain and officers—especially in the case of entirely new vessels—with particulars and data of the vessel's technical character, such as are now left to be found out by slow and sometimes bitter experience. Of these it may be sufficient to instance:—Stability, steadiness, trim, carrying capability, and steaming powers. Mr. William Denny, of Dumbarton, has recently publicly declared his firm's intention of supplying such particulars to the vessels built by them. It is to be hoped this worthy example may be extensively followed.

which rough weather or rank carelessness make always
imminent have come out of the ordeal with credit to the
constructors. One notable case may be instanced. The
Arizona, of the Guion Line, some time after being put on the
Atlantic service, while steaming at a speed of 14 knots, and
almost in mid-Atlantic, ran into an iceberg of gigantic dimensions, and notwithstanding that the force of the concussion
smashed her bows for a length of 20 feet into an unrecognisable
mass, she kept afloat, and reached a port of safety.

Where, as has already been indicated, there is such close
oversight and thorough supervision—where, indeed, the real
interests of every party honestly concerned lie so clearly in
the high qualities of construction—nothing short of such
results as the foregoing should be expected. The insurance
companies, on whom the burden (monetary at least) of loss
at sea ultimately falls, see it their interest to know that those
registration societies, on whom they rely for guarantee as to a
vessel's structural and general efficiency, are themselves efficient and trustworthy authorities. These societies, known as
Lloyd's, Liverpool Underwriters, and Bureau Veritas, Registries,
in spite of the dread as to business rivalry affecting injuriously
their standards of classification, have still a high criterion, and
enjoy the confidence of insurance societies and shipowners alike.

Shipowners themselves, notwithstanding some examples to
the contrary, are, and have always been, anxious and pains-
taking seekers after thoroughness; not merely mercenary
grubs, sacrificing considerations of safety to features promising
exemption from tonnage or other registration dues, and perhaps
the extinction of a rival. Some of the best British vessels,
notably those of the Cunard Line, are unclassed at the registries, but have been built under private survey. The well
known boast of the Cunard Company that not a single life
has been lost by mishap at sea during their long and extensive
service, is eloquent testimony to the care exercised in the
construction and management of ships. It is the practice of
some companies to effect classification in two, sometimes
three, separate registries, and the number of inspectors

PORTRAIT
AND
BIOGRAPHICAL NOTE.

———

BENJAMIN MARTELL.

Within the period covered by this review, this eminent inventor has introduced an instrument which enables soundings to be taken while vessels are going at full speed, at depths of 100 fathoms and under. The sounding line adopted is a fine steel wire, such as is used by pianoforte makers, which passes through the water with very little resistance, and can be sent to the bottom by a light weight or sinker, even when the ship is going full speed. Fastened to a short length of rope, near the sinker, there is a brass tube, in which is placed a glass tube two feet long, closed at one end and open at the other. This glass tube is coated inside with chromate of silver. As the sinker goes down, the air in the tube becomes compressed, and sea water rises up inside, the height to which it rises depending on the depth, from the surface, to which the glass tube goes down. As the sea water rises in the tube, the salt of the water acts on the chromate of silver and changes the colour from red to white; thus a mark is left on the glass tube showing the height to which the sea water rises, from which the actual depth may be at once measured by a prepared scale. By means of this sounding machine a ship can feel her way round a coast in a fog without reducing speed. In later instruments the inventor has devised another form of automatic gauge, which obviates the use of glass tubes, and is a decided improvement on the gauge here described.

The well-known Improved Mariner's Compass introduced by Sir W. Thomson enables the magnetism of the ship to be completely corrected instead of only approximately. This is attained by the use of several small needles instead of one or two large ones. The requisite steadiness of the compass card is obtained by means of an aluminium rim suspended round the edge of the card. The extreme lightness of the card reduces greatly the wear of the needle point supporting the compass. Along with the compass the inventor supplies an azimuth mirror which greatly facilitates observations either on a point of land or on a star, the whole invention proving from experience an almost indispensable item of outfit for well-appointed vessels.

The care and ingenuity expended on the question of ship safety must not, however, be measured simply by the amount of attention and skill exercised in constructing and outfitting vessels of the common type. The question has very naturally occasioned many distinct novelties in ship design. Some of these have been directly designed to secure safety, but the greater number have aimed at combining with safety the other qualities of speed and comfort; as in the instances given in the previous chapter. The success attained in practice, it need scarcely be said, has hitherto been but partial.

The problem of rendering ships absolutely unsinkable has, from very early times, received attention from many concerned in shipbuilding and navigation. Propositions and trials have been made from time to time, without as yet any very marked success attending any of them. Various plans have been submitted for safety-ships, the general principle of which consists in forming the ship into two or more distinct and entire portions, and in the event of one sustaining damage by collision or otherwise, those remaining to be disconnected and sent adrift—presumably with all passengers on board.

Other life-saving devices, while interfering somewhat with original structure, have simply been intended to use or modify existing features or material on board ship. Two of these which have received attention from the Scientific Societies may be shortly described as examples of the class of devices referred to. One was the proposition of Mr. Jolly, M.A., of the Royal Navy, laid before the Institution of Naval Architects in 1874; the other being that of Mr. Gadd, submitted to the Manchester Mechanical Society in 1879. Mr Jolly's proposal was to construct what he felicitously termed the "ark saloon," an erection on the upper deck, and resembling very closely an ordinary deck-house, but instead of being built permanently on the vessel, it was to be an independent structure capable of being readily disconnected, and "while answering all the purposes of accommodation found in ordinary deck-houses, to have within it hidden resources capable of converting it when afloat into a perfectly navigable vessel." Mr. Gadd's proposal was to form the upper portion of the bulwarks of ships of loose sections 12-ft. long, composed chiefly of hollow, thin metallic tubes. These sections when immersed in the water would form so many pontoons, and would be provided with cords and loops along their sides, and in the event of the ship going down would be lifted out of their place by the action of the water. Objections on economical grounds to Mr. Jolly's scheme, fully pointed out by members of the Institute, apply almost equally to the proposal of Mr. Gadd. The expense involved in their application would far outbalance in the eyes of the shipowner the possible service they could render. No provision was made by Mr Jolly for launching his ark saloon, thereby limiting its use to cases of foundering; and even in event of this, the "ark" was only to be so in name until the good ship should "go under," and leave the saloon serenely floating—presumably with all souls inside. The difficulty in Mr. Gadd's proposal, of at once making the bulwarks easily floatable and structurally efficient for the resistance of heavy seas, seriously detracted from its feasibility.

It would be a somewhat heavy task to make adequate note of all the varied proposals and patented inventions for the preservation of life at sea. Some of these, as in the foregoing instances, are proposals affecting structural features; but others, and by far the most numerous, are simply adjuncts to the vessel. Ingenuity has been specially directed of late towards bringing into efficient requisition, in event of impending shipwreck, the commonest items of a ship's outfit. This has been abundantly evidenced in the several naval exhibitions held within the past three years in various parts of the country. Firms whose work lies in cork and India-rubber manufactures have there exhibited in great profusion various forms of life-belts, life-buoys, life-saving mattresses and pillows, and life-saving dresses. Others, availing themselves of larger items, have shown life-saving adaptations of deck-seats, deck-houses, and bulwarks made into the form of life rafts. Not a few of these devices have received adoption in our passenger-carrying steamships, and their more general use

—especially if accompanied by proper knowledge of how they may best be taken advantage of—would materially help to rob shipwreck of some of its terrors at least, if not of its dire fatali- ties. It has been urged in this connection—and the plea is eminently reasonable—that Parliament should invest the Board of Trade with proper powers—if that Body is not already vested with all that is requisite—to take the matter of life-saving appliances thoroughly and practically in hand, and by means of experiments in all kinds of weather to determine which are the best means of saving life under different conditions. Having done this, also to draw up rules for the proper stowage and use of such appliances on board ship, and to see that such rules are strictly observed, and that no vessel be permitted to go to sea which is not so equipped.

The development in the size of steamships not only affects the quality of safety, but also in various ways the element of comfort at sea. The greater length, for instance, is calculated to neutralise the longitudinal oscillation, the effects of which are so often fatal to the comfort of passengers. Again, the great length affords an advantage in the way of allowing better state-room accommodation; all the rooms, or a larger proportion of them, being next the vessel's side, and conse- quently more airy and better lighted. It is not, however, in the increased length so much as in the development of all three dimensions, and especially in the increased ratio of breadth to length, that modern types of steamships are enhanced in the qualities of safety and comfort. Mistaken or imperfect notions as to the ratio most desirable for speed, have kept in perpetuity types of steamers which the fuller light of modern scientific investigation has shown to be undesirable. Great beam is now believed to be not incompatible with great speed, and even apart from questions of speed the advantages accruing from breadth are better appreciated.

As an illustration of this movement, one of the more recent of the many transatlantic mail steamships may be instanced.

In the *Aurania*, of the Cunard Company, the proportions—although perhaps only in the line along which modern professional ideas tend—are certainly in advance of the general practice with regard to vessels of her great size. The dimensions of the *Servia*, the *Alaska*, and the *City of Rome*—three vessels comparable with the *Aurania* as constituting the largest merchant vessels afloat—all give a proportion of 10 beams to the length. The *Aurania's* dimensions—470 feet by 57 feet by 39 feet—show her to have only about 8¼ beams to length. The success of the older type of vessel having proportions somewhat similar to this "modern instance" has in no material sense been eclipsed by the narrow types which subsequently for so long prevailed. Availing themselves of that freedom which independence of the registration societies yield—their vessels not being "classed"—the Cunard Coy. determined to adopt the old-time proportions. The step has been justified, in so far as affected by the matter of speed, the powerful vessel, at her trials on the Clyde, having attained a mean speed of 17¾ knots, or 20½ statute miles, per hour. The stable qualities due to the great breadth of the *Aurania* has in actual service further confirmed the wisdom of the change. The magnificent vessels presently building on the Clyde for the Cunard Coy., though between 20 and 30 feet longer, are the same breadth as the *Aurania, i.e.,* 57 feet. This is accounted for by the fact that the breadth of beam fixed for the *Aurania* was the largest amount permissable, having regard to the breadth of entrance of the largest dock in New York. This *en passant* is worthy of notice as giving colourable justification to the complaints sometimes made that civil engineers are urged to progress in dock accommodation only by shipbuilders treading on their heels.

Coincident with the changes made in the dimensions and structure of vessels, there are numerous features of enhanced comfort for passengers and crew which are deserving of notice. Notably is this manifest in the arrangement of saloons and state-rooms—their appointment, lighting, and ventilation. The character of steamships for the great ocean highways in this respect is above and beyond anything which Board of Trade enactments seek to secure. The amount of spirited competition itself on those services, acts as an efficient promoter of excellence in design and equipment.

It is now the prevailing fashion to appropriate that part of a steamer just before the engine and boiler hatchways for the principal saloon and first-class berthing, and it has so many advantages over the old plan of locating these apartments in the poop or after extremity of the vessel that its adoption in large steamers of the passenger carrying trade has become all but general. Some of these advantages may be briefly enumerated. They are:—ampler and airier saloon space: the plumbness of the vessel's sides permitting a saloon completely athwartship, which is scarcely practicable in the conventional situation aft, because of the curvature of sides; increased facilities for ventilation; purer air; freedom from the noise and vibration caused by propeller; comparative immunity from the effects of "pitching" or longitudinal oscillation.

Nothing, perhaps, in connection with improved saloon accommodation strikes one so much as the increased height between decks now prevalent. While from six-and-a-half to seven-and-a-half feet was considered sufficient some years ago, it is now the practice in first-class steamers to make the height as much as from eight-and-a-half to nine-and-a-half feet. The feeling of spaciousness this change contributes to the saloons, as well as the scope it yields for architectural treatment of the walls, are not the least gratifying results of the improvement. How much the latter result has been taken advantage of in our modern passenger steamships need scarcely be told, as their architectural and decorative character is often and eloquently enlarged upon by delighted voyagers.

A noteworthy feature in improved saloon accommodation is the provision of music rooms or social halls, which are usually situated above the dining saloons, and connected or made one therewith by means of light and ventilation wells placed in the centre. The size and ornamentation of these, and the light and air they are the means of admitting, contribute in a very

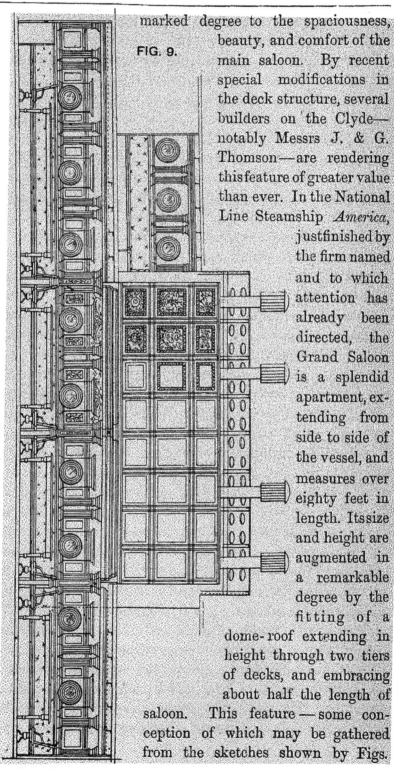

FIG. 9.

LONGITUDINAL SECTION OF GRAND SALOON IN S.S. *America*, SHOWING DOME-ROOF.

marked degree to the spaciousness, beauty, and comfort of the main saloon. By recent special modifications in the deck structure, several builders on the Clyde—notably Messrs J. & G. Thomson—are rendering this feature of greater value than ever. In the National Line Steamship *America*, just finished by the firm named and to which attention has already been directed, the Grand Saloon is a splendid apartment, extending from side to side of the vessel, and measures over eighty feet in length. Its size and height are augmented in a remarkable degree by the fitting of a dome-roof extending in height through two tiers of decks, and embracing about half the length of saloon. This feature—some conception of which may be gathered from the sketches shown by Figs.

9 and 10, is altogether free of athwartship beams, and practically gives to the saloon a clear height of 1 8 feet. The crown of the dome is formed of beautifully-executed stained glass, finished round its base in a richly coloured frieze formed

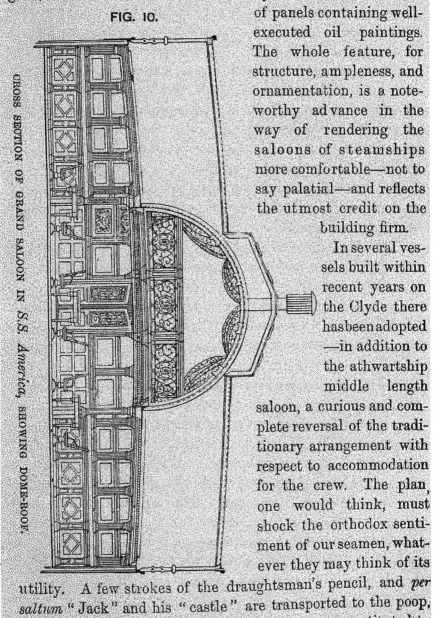

FIG. 10.

CROSS SECTION OF GRAND SALOON IN S.S. *America*, SHOWING DOME-ROOF.

of panels containing well-executed oil paintings. The whole feature, for structure, ampleness, and ornamentation, is a noteworthy advance in the way of rendering the saloons of steamships more comfortable—not to say palatial—and reflects the utmost credit on the building firm.

In several vessels built within recent years on the Clyde there hasbeen adopted —in addition to the athwartship middle length saloon, a curious and complete reversal of the traditionary arrangement with respect to accommodation for the crew. The plan, one would think, must shock the orthodox sentiment of our seamen, whatever they may think of its utility. A few strokes of the draughtsman's pencil, and *per saltum* "Jack" and his "castle" are transported to the poop, and the precincts so long sacred to his use are prostituted to the lounge and the tobacco pipe of the pampered "land-lubber" —*i.e.*, they form a luxuriant smoking saloon for passengers.

Of the multifarious ways in which modern invention and
skill are laid under contribution to the end that voyagers shall
have the maximum of safety and comfort on board ship, the
system of electric lighting now so extensively adopted is not
the least noteworthy. It is only about three years ago since
the application of the incandescent form of electric lamp on
board ship was first tried. The success of the system and its
rapid extension during the subsequent period has been remarkable, and is a matter upon which electricians, shipowners, and
sea voyagers are alike to be congratulated. In every well-
appointed passenger ship for ocean service, the electric light
has already supplanted the former method of lighting the
saloons, state-rooms, and machinery spaces, by means of oil
lamps, which has so often proved a fruitful source of annoyance to passengers and crew, if not, indeed, of positive danger
to the vessel herself.

The advantages of the change are such as constitute the
electric light an invaluable acquisition on board every modern
passenger steamship. The light gives off very little heat, there
is no smell, no products of combustion to produce headaches
and sickness. No matches are required, and the danger from
fire is absolutely reduced to a minimum. The light requires
little or no attention on the part of stewards, for it is only
requisite that a man be sent round once a day to see whether
any of the lamps require renewal, and the renewal of a lamp
is performed as simply as trimming the wick of an oil lamp or
placing a fresh candle into a candlestick. The danger, annoyance and time, formerly spent in storing up and dealing out
large quantities of paraffin or other oils, are completely obviated.
The lamps are as easily subject to the control of the passenger
as ordinary gas jets. Instead of the flickering and somewhat
clumsy oil lamps, the electric system presents, encased in neat,
tiny, glass globes, a steady, mellow white light, the adaptability
of which to any conceivable position or design is one of its most
beautiful properties. The artistic grouping of the electric
incandescent lamps, and their combination with the architectural features of saloons, are matters to which the forms adopted

for the best known lamps—the Edison & Swan types—specially lend themselves. A single Edison lamp is shown by Fig. 11.

The work in electric-lighting on board ship for the year 1883 shows how firmly the electric system has become established as the only system for first-class passenger vessels. The report of the Edison & Swan United Companies embraces the work on thirty-one vessels, including three Indian troopships (and four more on order), four vessels for the Clan Line, one for the Peninsular and Oriental Company, one for the Union Steamship Company, three for the Cunard Company, three for the British-India Steam Navigation Company, three for the New Zealand Shipping Company, and so on. The list of Messrs Siemens Brothers amounts to twenty high-class vessels, including the *Arizona*, the *Servia*, the *Aurania*, the *City of Rome*, the *City of Chicago*, the *Austral*, the *Germanic*, and the *Massilia*. These two firms thus give fifty-one vessels, and adding those entrusted to outsiders—four in all—affords a total of fifty-five, representing an aggregate of not less than 11,000 incandescent lamps.

FIG. II.

EDISON LAMP.

The application of the electric light on board ship to the purposes of signalling, as a substitute for the ordinary system of oil lanterns, has been fully shown in theory and already partially effected in practice, but its development in this direction is necessarily retarded by considerations which do not affect its use in the interior of vessels. Vessels traversing the ocean in darkness are necessarily dependent one on the other for the means of knowing their proximity, and as the electric light much exceeds in power and brilliancy that of oil lanterns, it would have the effect of eclipsing the latter even within a large radius. The adoption of the electric light for this very important purpose would, therefore, have to be pretty much a simultaneous and general movement throughout the ships of the various companies, if not of the various nations. Apart from such considerations, however, other objections have been instanced to the appropriation of the electric light for this purpose. Difficulty, it is said, has been experienced in distinguishing the colours pertaining to the port and starboard side-lights, and fears are entertained regarding the liability of the light, or the machinery employed in generating the current, to suddenly fail in its action. Few of the objections named, of course, amount to very serious obstacles, and as the system is yet so much in its infancy, it may well happen that a few years will witness all that is here foreshadowed.

Short of this universal and complete appropriation of the electric light for signalling, however, it has been introduced with gratifying results in mercantile steamers for various important purposes—*e.g.,* for lighting up the decks and surrounding wharfage during the work of loading or disembarking cargo; for projecting a flood of light ahead of a vessel's course where navigation is difficult, and when danger in the shape of rocks or icebergs is imminent. The employment of the light in the way last named has been specially extended in the case of vessels intended for naval warfare. By its powerful aid the position and tactics of the enemy, the configuration of forts about to be assailed, or the nature of the land where it is proposed to disembark, can all be revealed, with a minuteness almost as perfect as that due to the light of day.

Another feature on board ship affecting most intimately the well-being and comfort of passengers—too often, indeed, the safety of the ship itself—is that of ventilation. The thorough and efficient ventilation of ships is a feature which only during very recent times has received from shipowners and shipbuilders the amount of attention it deserves. The inadequacy of the methods of ventilation existing in emigrant ships, and as applied to holds for the ventilation of cargoes, engaged public attention very considerably a few years ago. The explosion on board the *Doterel*, with other like casualties, resulted in the appointment of a Royal Commission to inquire into the ventilation of ships. The prominence thus given to the subject and the experience then gained, have been fruitful of increased regard for efficiency in ship ventilation. In the absence for such a long time, however, of any system capable of universal application and having at once the merits of efficiency and cheapness, shipowners have adhered to old-fashioned, unscientific, and ineffective methods long after the invention of improved systems, one or other of which would have well repaid adoption.

In ways and to an extent which may perhaps have been made evident in the previous pages, the introduction of the electric light is of itself greatly advantageous in this connection. One striking peculiarity of the change perhaps requires more explicit statement. This is the curious fact—patent enough to all who know anything of the properties of the incandescent light—that what is the very life of oil or other lights, is to it, certain death. The element thus vitally concerned is, of course, oxygen; and it need not be more than hinted that in existing so entirely without this element—at all times a great desideratum in passenger ships—the electric light is a vast benefactor to all who " go down to the sea in ships."

Many highly-improved methods of ventilation are now open to the shipowner; the number of patented systems in use or awaiting adoption being adequate testimony to the widespread attention bestowed upon the subject. These divide themselves into two general classes:—firstly, systems which aim at providing an efficient self-acting series of ventilating pipes in which the natural current or that induced by the vessel's own speed through the atmosphere, is the only force utilised; and secondly, those in which machines driven by steam power are employed to produce fresh currents or extract vitiated atmosphere.

Various forms of ventilators, belonging to the first-named class, have been introduced into many ocean-going passenger vessels within recent years, the result being a considerable improvement in the sanitary condition of the more confined portions of vessels. One of the most approved of these, receiving specially extended adoption, amounting as it does to a highly perfected system, may be noticed a little in detail. This is the form of ventilator patented and introduced by Messrs R. Boyle & Son, the well known ventilating engineers of

London and Glasgow, consisting of upcast and downcast shafts
fixed above deck, communicating with the interior of vessel
by a system of piping led to the various compartments. The
upcast, or "air pump" ventilator, as the patentees term it,

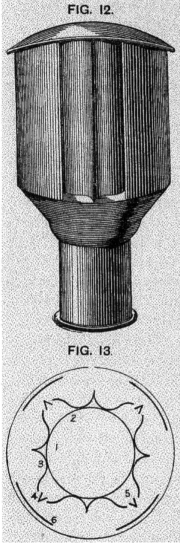

FIG. 12.

FIG. 13.

consists of a fixed head having an
ingenious arrangement of louvre
webs, whereby the wind impinging
upon it from any direction, creates
a current and exhausts the air from
the cylinder of which the head is
part, the foul air from below immediately ascending to supply the
place of the air extracted. A continuous and powerful upward current is thus induced, and the head
is so devised as to effectually prevent
down-draught or the inlet of water.
The elevation and plan of this ventilator is shown by Figs. 12 and
13. In Fig. 13, 1 represents
cylindrical chamber communicating
with shaft below; 2, deep lip to prevent the possibility of water passing
into cylinder and down the shaft;
3, curved plates to deflect and com-
press the air over outlet openings or
slits; 4, creates an induced current
and exhausts the air from the
cylinder; 5, radial plates to deflect
air off centre of slits; 6, curved
baffle plate or guard, to concentrate
the current, and prevent the wind
blowing through the slits opposite. The downcast ventilator,
though necessarily more simple, is arranged, by means of similar
louvered webs to prevent any water passing below, lodging it
on the open deck instead. By means of up and downcast ventilators of this type, it is possible to have the ventilation

going on between decks without interruption when there is a
storm blowing and seas sweeping the deck, whereas under
ordinary conditions, and in similar weather, everything would
be battened down and the ventilation *nil.* The inventors, of
course, are able to point to other advantages possessed by
these ventilators, but the above are the salient features, which
have won for their system marked recognition and pretty wide
adoption. As evidencing its efficiency, it may be stated that
Messrs Boyle's system was awarded the "Burt" prize of £50,
offered for international competition by the Shipwrights' Coy.
of London in 1882 for the best system of ship ventilation.

Having regard to the great importance of first providing
means whereby foul air may be extracted from compartments
rather than first attempting to put fresh air in—at least by
other than mechanical means—it has become the practice with
several steamship companies to fit a series of pipes from the
rooms throughout the 'tween decks all leading into a common
main, carrying this main into the boiler funnel, and thus
utilising the powerful draught existing there when the vessel
is under way. The efficiency of this method is all that could
be wished, but its action is necessarily impaired when the
vessel is in port and the boilers not in use. For steamships
having long runs its value is very considerable; but in steamers
having short passages and long port delays its merits are not so
pronounced, and it is, of course, of no account when sailing
ships are concerned.

Two systems of ventilation much alike in principle and
equally applicable to the steamer and sailing ship may be
shortly referred to. One is the Norton Ventilator, in which
the dipping motion of vessels is utilised in effecting their own
ventilation; the action in ocean-going vessels, of course, being
continuous and automatic. Two cylinders, closed at the upper
ends, are placed on each side of the stern post at such a distance
as not to interfere in any way with the action of the rudder,
and sufficiently close under the stern to be well out of harm's
way. As the vessel rises the water drops in these cylinders,
which are partly submerged, and in its fall causes a vacuum,

to fill which the air is drawn from all parts of the ship. The sinking motion of the vessel again fills the cylinders and forces the foul air collected, through the discharge pipes. The ventilator admits also of being actuated by steam or other power on board steamships, the exhausting and forcing device in this case consisting of a water bell or air chamber to which a vertical reciprocating motion is imparted by a beam or other attachment operated upon by the mechanical power adopted. The other method referred to is that now being pretty extensively introduced by Messrs Mosses and Mitchell, of London. It consists of two small cylinders, placed on either side of a ship, in-board, and connected by a pipe. The cylinders are partly filled with water, and, as the vessel rolls, the water rushes from the elevated to the depressed side of the ship, from one cylinder to the other, and, by creating a vacuum, draws up the foul air from between decks, or out of the hold, by pipes leading below. The air which is pumped up by this self-acting process goes out through a discharge pipe over the side, and such is the force of its exit that it serves to blow a fog-horn when required. The cylinders can be placed so as to be worked by the pitching as well as the rolling of the vessel, and there is always a sufficient movement of the water to keep these pumps in action.

Systems of the other class—those involving the aid of mechanical power—are as much available as the automatic systems, but the greater expense of fitting, maintaining, and working them are considerations, apart from the question of their greater efficiency, which stand in the way of their general adoption. In vessels chiefly intended for passenger or emigrant carrying, artificial ventilation by mechanical means has been provided, and the practice is greatly on the increase, but systems in which natural agents are more largely brought into requisition have advantages which appeal most effectually to ship owners in general.

In several modern steamships engaged in cargo and passenger service, hydraulic machinery designed to take the place of the usual deck steam equipment has recently been introduced with great advantage. This embraces machinery used for steering the vessel, loading and discharging cargo, heaving anchors; for performing, indeed, all the work on board excepting that of propulsion. From experience of the well proved utility and durability of hydraulic power on shore, it seems quite a natural consequence that it should take its place on board ship. Indeed, the system has so many advantages both from the point of view of the passenger and of the steamship owner, the wonder is that its introduction has been so long delayed. Its perfect noiselessness, as compared with the rattling, hissing, steam machinery now in vogue, is an advantage which will appeal strongly to the sea voyager. The great speed of the system, as well as the absence of jar and noise, the reduction in wear and tear, and the obviating of well-known disadvantages incidental to steam pipes, are merits of the system which are bound to appeal to the steamship owner.

It has been well pointed out by Mr. A. Betts Brown, of Edinburgh, the patentee and manufacturer of this class of machinery, in a paper read by him before a recent meeting of the Institution of Naval Architects that—"With all the noise of steam engines at work on deck, running at piston speeds of as much as 1000 feet per minute, the cargo is lifted from the hold at a rate of only from one to two feet per second, which cannot be considered as keeping pace with the general progress made in other departments of steamship economy. In short, vast sums are spent on fuel to gain half a knot extra speed on a passage, while hours may be wasted in port in consequence of the primitive nature of the present system of deck machinery for discharging cargo." Previous to 1880, Mr. Brown had supplied and fitted hydraulic machinery on board the paddle-steamer *Cosmos*, built by Messrs A. & J. Inglis, of Glasgow, intended for South American river service, but it was only in that year that he had an opportunity of fitting a large ocean-trading steamship with the system. This was the *Quetta*, built by Messrs Denny, Dumbarton, to whom, with the managers of the British India Association Steam Navigation Company, who own the vessel, Mr. Brown ascribes credit for the opportunity

afforded him of fitting his firm's system on a complete scale.
The *Quetta* is 380-ft. in length, 40-ft. breadth, depth of hold
29-ft., and 3,302 tons gross, and is fitted with a complete system
of hydraulic machinery performing the following functions:—
Steering, heaving the anchor, warping by capstans fore and aft,
taking in and discharging cargo, lowering the derricks to clear
cargo over side, hoisting ashes, reversing main engines, and
shutting tunnel water-tight door in engine-room.   For detailed
descriptions of these various appliances, the reader is referred
to the before-mentioned paper.   The most for which space is here
available is a very general outline of the principle on which
they are supplied with motive power.   The prime mover consists of a pair of compound surface-condensing pumping engines
of 100 indicated horse-power, situated in the engine-room of
the vessel.   These engines pump water (or in winter non-
freezing fluid) from a tank into a steam accumulator.   The
pumping engines are started and stopped by the falling or
rising of the steam piston in the accumulator; and since the
piston falls when the hydraulic power is being utilised, and
rises to its former level when the power is not in use, it follows
that the apparatus is perfectly automatic.   Once started, it
does not require the supervision of an engineer, and it main-
tains a steady pressure of 800-lbs. per square inch in the
hydraulic mains or pressure pipes.   These are carried up from
the engine-room, and extend fore and aft the ship.   Alongside the pressure main a similar return main is laid, which
discharges into the tank.   From the pressure mains branches
are connected to the various hydraulic machines.   After having
done its work, the water is discharged into the return mains,
being thus used over and over again.   The experience obtained
in the working of the *Quetta* shows that a donkey boiler of
the usual size, just sufficient for steam winches, enables the
cargo to be discharged in half the time: in other words, does
double the work on a given coal consumpt with compound surface-condensing pumping engine, and the hydraulic system.

The advantages of hydraulic machinery have been thus
summarised:—A pair of engines in one place do, with no

noise and half the consumption of fuel, the work usually performed by perhaps a dozen donkey engines, while about £30 or £40 a voyage is saved in wear and tear. The increase of speed obtained in loading and discharging cargo practically ensures a quicker voyage The rapidly working machinery necessitates double gangs of men in the hold; but though the hands are more numerous they are paid for a shorter time, and the cost of labour per ton of cargo is thus less than usual. The prime outlay is considerably greater than under the ordinary system, but it is calculated that in at least three years the extra expense will have been saved.

Notwithstanding the considerable increase in cost (more than double that of steam equipment) of the hydraulic system, the British India Association have seen their way to fit the succeeding steamers they have built, similarly to the *Quetta*, namely, the *Bulimba, Waroonga,* and *Manora*, the two intervening ships having their emigrant quarters ventilated by fans driven by hydraulic engines, as well as the usual deck equipment. In addition to the above, there have already been nine other steamers fitted successfully by Mr. Brown's firm with hydraulic machinery—including the Union Steamship Coy.'s *Tartar*, of 4340 tons—and there is every prospect now of its taking the place of the noisy steam machinery in at least our most important passenger lines.

The regard which is had to comfort and luxury in modern passenger steamers has manifested itself—like the attention devoted to swiftness and safety—in various propositions and designs of a more or less novel kind. These, indeed, have very often consisted of designs embracing the whole of the qualities named ; comfort and luxury being coincident with the more important properties of speed and safety already noticed ; but not a few propositions and actual undertakings have consisted of vessels in which comfort has largely been the dominant and regulating condition of design. This subject receives happier illustration from the history of steam service between England

and France, than perhaps from any other service that could be instanced. The thought and speech expended on " an efficient Channel service" at the meetings of the various societies concerned with shipbuilding and marine engineering, and the space devoted to the subject in the technical journals, has been no more than commensurate with the number and variety of projects for its accomplishment, submitted from time to time. Many of the schemes have not been quite of a marine character, and these, of course, lie beyond the province of the present review ; but so far as ships are concerned, it is interesting to note to what extent comfort has been the dominant and regulating condition in the designs. In the *Castalia* and the *Calais-Douvres*, employed in Channel service, features of considerable novelty—notably the double hulls—were adopted, and it was to the desire for increased comfort as much as speed that their introduction was owing.

In the steamer *Bessemer*, however, built at Hull in 1875, this subject finds happiest illustration. This steamer, which involved some very interesting and novel problems in shipbuilding—in which the matters of propulsion and steering were largely concerned—was designed for the special purpose of practically testing an invention of Mr Henry Bessemer's, having as its object the alleviation of the evils of sea-sickness. Mr (now Sir Henry) Bessemer's invention, as applied in this case, consisted of a saloon supported on longitudinal pivots, which was to be made unsusceptible to transverse oscillation by the application to it of machinery wrought by hydraulic power. It was the intention of the eminent inventor to have applied this system to the correction of longitudinal as well as transverse oscillation, but on considering that the steamer was to be of large dimensions and performing a service in comparatively small waves it was thought desirable to limit its application to transverse motion, at the same time having regard to the longitudinal motion by reducing the height of the vessel for a distance of 50-ft. at each end, thereby inducing depression at the extremities, through the vessel's not rising to, but being overswept by, the waves.

Although an influential company was formed to work the *Bessemer* and other vessels embodying her novel features, which it was thought might follow, she was virtually abandoned after a single trial across the Channel. Her failure was assumed without exhaustive and conclusive trials being made of the many novelties embodied in her construction, some of which were obviously of an experimental character. This is the more to be regretted because of the beneficent issues involved in the project, and also in some degree because of the extent to which the faith of some intrepid and experienced men was pledged to its success. Nevertheless, it was always a matter of grave doubt, even when the fullest measure of mechanical success was allowed for, whether the idea of the pivoted saloon was calculated to secure that immunity from the effects of ship motion in a sea-way, for which the celebrated patentee felt induced to hope.

It is maintained by many who profess to have given the subject attention, that sea-sickness in its most virulent forms, and in the majority of instances, is less attributable to the transverse and longitudinal oscillations—known respectively, as the "rolling" and "pitching" motions—than to the vertical movement termed "dipping," which in its descent from the summit of one wave until upborne by the wave next following, the vessel undergoes. Now, this is a condition for which, in the Bessemer project, there was no provision, nor indeed well can be under any circumstances, save in the simple but costly expedient of adding to the dimensions or bulk of vessels, irrespective of form. The Czar of Russia's yacht *Livadia*, built some years ago, exemplified in her extraordinary dimensions and great bulk the truth of such reasoning. The actual rolling and pitching of this remarkable vessel, as observed in the height of a gale in the Bay of Biscay, and in the midst of very heavy seas, was exceeding small. This never exceeded four degrees for the single roll, or seven degrees for the double roll, nor beyond five degrees for the forward pitch, or nine degrees for the double pitch, so to speak. This horizontal steadiness appeared to experts, who were on board at the time, most

remarkable, and Sir E. J. Reed, in a communication to the
Times, commented amongst other things on the agreeableness
of the contrast the voyage on the *Livadia* afforded, with his
experience of voyaging at sea in ordinary ships.

After all, it must be acknowledged that attempts hitherto
made to obviate the evils of sea sickness by novelty in design
fall very far short of attaining the beneficent results sought
after. The *Bessemer*, the *Livadia*, the *Calais-Douvres*, and other
unique craft primarily conceived with regard to this end, are
now, it would seem, exemplifying in their latter fate the futility
of the endeavour.· Such attempts, however ill-advised they
may possibly appear in the light of the knowledge their very
failure or their partial successes yield, have still their credit-
able and praiseworthy aspects. The spirit which has prompted
some of them is not wholly one of money-making, and their
histories enrich the general fund of experience far more than
libraries of untried theories. Shipowners are too ready to shut
their minds against everything which seeks the *acme* of comfort and safety by other means than those which guarantee
economical success, or those which consist in increasing the
size and power, and enhancing the accommodation of conventional types of vessels. These novelties and innovations, on
the other hand, represent more of the intrepidity essential to
genuine advancement than is forthcoming in a thousand merchant ships of the conventional type.

Happily the need for such enterprise as is involved in at
once departing from tried types, has within recent years been
largely, if not altogether, obviated, through improved procedure
in the work of design. The more thoroughly analytic process
of investigation and experiment now in vogue, greatly curtails
the number of novelties introduced, or which reach the constructive stage. Many present-day projects never get beyond
the "paper stage," which in times not so far distant would
have spelled out "failure" to the very last letter. Since the
system of model experiment has begun to be practised in a
reliable manner, and since theoretical prediction generally has
become better appreciated, over-sanguine inventors have been

spared the penalties of failure in actual practice, and ingenuity has been reclaimed or warned away from channels that would inevitably have proved chimerical.

List of Papers bearing on the safety and comfort of modern steamships, to which readers desiring fuller acquaintance with the *technique* and details of the subjects are referred:—

ON THE NECESSITY OF FITTING PASSENGER SHIPS WITH SUFFICIENT WATERTIGHT BULKHEADS, by Mr Lawrance Hill : Trans. Inst. N.A., vol. xiv., 1873.

ON WATER AND FIRE-TIGHT COMPARTMENTS IN SHIPS, by Mr Thomas May : Trans. Inst. N.A., vol. xiv., 1873.

ON CAUSES OF UNSEAWORTHINESS IN MERCHANT STEAMERS, by Mr Benjamin Martell : Trans. Inst. N.A., vol. xxi., 1880.

ON MODERN MERCHANT STEAMERS, by Mr James Dunn : Trans. Inst. Naval Architects. vol. xxiii., 1882.

ON BULKHEADS, by Mr James Dunn : Trans. Inst. N.A., vol. xxiv., 1883.

ON PUMPING AND VENTILATING ARRANGEMENTS, by Mr Thomas Morley : Trans. Inst. N.A., vol. xvii., 1876.

ON SIR WM. THOMSON'S NAVIGATIONAL SOUNDING MACHINE, by Mr P. M. Swan : Trans. Inst. N.A., vol. xx., 1879.

ON STEAMSHIPS FOR THE CHANNEL SERVICE, by Mr John Grantham : Trans. Inst. N.A., vol. xiv., 1873.

ON CHANNEL STEAMERS, by Mr John Dudgeon : Trans. Inst. N.A., vol. xiv., 1873.

ON HIGH-SPEED CHANNEL STEAMERS, by Mr H. Bowlby Willson : Trans. Inst, N.A., vol. xv., 1874.

ON THE ARK SALOON, OR THE UTILISATION OF DECKHOUSES FOR SAVING LIFE IN SHIPWRECK, by Rev. W. R. Jolley, R.N. : Trans. Inst. N.A., vol. xv., 1874.

ON THE BESSEMER STEAMSHIP, by Mr E. J. Reed : Trans. Inst. N.A., vol. xvi., 1875,

ON THE BESSEMER CHANNEL STEAMER : Naval Science, edited by Mr E. J. Reed, 1873.

ON ELECTRIC LIGHTING FOR SHIPS AND MINES, by Mr Andw. Jamieson : Trans. Inst. Engineers and Shipbuilders, vol xxv., 1881-82.

ELECTRICITY ON THE STEAMSHIP (Series of Papers) : the "Steamship," vol. I., 1883,

ON THE VENTILATION OF MERCHANT SHIPS, by Mr Jas. Webb : Trans. Inst. N.A., vol. xxv., 1884.

ON THE COMPARATIVE SAFETY OF WELL-DECKED STEAMERS, by Mr Thos. Phillips : Trans. Inst. N.A., vol. xxv., 1884.

ON THE APPLICATION OF HYDRAULIC MACHINERY TO THE LOADING, DISCHARGING, STEERING, AND WORKING OF STEAMSHIPS, by Mr A. B. Brown : Trans. Inst. N.A., vol. xxv., 1884;

CHAPTER IV.

PROGRESS IN THE SCIENCE OF SHIPBUILDING.

THE appreciation and employment of scientific method and analysis in designing and building ships have at no previous time been greater than they are at present. This is already yielding benefits and ensuring successes which only a few years ago would have remained ungathered and unachieved, or at best would only have been attained after wasteful expenditure of money, time, and skill, if not the sacrifice of human life. Not so long ago endeavours were seldom made to extract lessons of general value from particular occurrences, there being a disposition prevalent to accept facts without accounting for them— "to rejoice in a success and regard a failure as irreparable"— the outcome, it may at once be said, of indifference, false ideas of economy, and of a limited conception of the part scientific methods should play in successful shipbuilding.

Particular occurrences within recent years have without doubt played a large part in bringing about this more general and intelligent appreciation of such matters. Some maintain, indeed, that it is only under pressure of circumstances that anything like proper regard for fundamental principles has obtained hold among mercantile shipbuilders. This remissness, even admitting it to be true, is the more natural and excusable in private commercial concerns, when it is considered that the bulk of progress made, even in Admiralty quarters—where ships take several years each to build, and there is more time for scientific investigation and experiment than is possible in mercantile work—is more attributable to the awakenings which have followed upon great disasters than to the natural improve-ment of ordinary practice. The terrible loss of the *Captain* in

September, 1870, for example, by which 500 lives were sacrificed, led to a fuller recognition of the necessity for exact experiment and calculation to determine thoroughly the conditions of stability for war vessels; and many warships then under construction at the dockyards—particularly those of the low-freeboard type—were altered in consequence, for the purpose of adding to their safety. The capsizing of the *Eurydice* off the Isle of Wight in March, 1878; the mysterious and mournful loss of her sister ship the *Atalanta* in 1880; the explosion on board the *Thunderer* in 1876, by which 45 lives were lost, and the still more calamitous case of the *Doterel* in April, 1881, by which the ship and 148 lives were destroyed, are all instances of calamity, the causes of which have formed the subject of official inquiry, all in their turn teaching important lessons and yielding subsequent benefits not easily calculable.

Recent occurrences of a very calamitous nature in connection with merchant ships—some of which will be more explicitly referred to further on—have been attended with similarly mournful, but, it may be added, with similarly beneficial results. These disasters and the resulting inquiries have shown pretty conclusively that the knowledge of a vessel's stability and other vital qualities possessed by ship's officers is often meagre and erroneous; and that far too little attention is usually paid to a vessel's technical qualities by shipowners or their advisers. They have also tended to prove that exact knowledge of the principles of ship design, and observance of scientific method in their construction, are not yet sufficiently prevalent or thorough in mercantile shipyards.

Progress in the pure science of naval architecture, as distinguished from the practical application of scientific rules and principles to shipbuilding, is a great and complex subject, and one which it would be impossible to do full justice to here. Before attempting to treat upon these matters as concerned with the period covered by this review, it may be instructive to trace briefly the progress made in the past, and take note of the agencies through which such progress has been effected. In this undertaking, concerned as it is with matters relating to a period prior to that with which the present work chiefly deals, the author has availed himself to some extent of already published works traversing the same ground. As having afforded the needful assistance in this connection, and as being a source to which readers may turn for fuller information, reference may here be made to an article in the *Westminster Review* of January, 1881, on "The Progress of Shipbuilding in England." This article, though unsigned, is from the pen of Mr. W. H. White, late Chief Constructor of the Navy, and author of the well known " Manual of Naval Architecture." It furnishes an appreciative and concise account of the literature and the educational agencies connected with the theory of naval architecture, and sketches the influence of science on practice, and *vice versa* in the profession since the beginning of the present century.

As has already been indicated, the period during which scientific knowledge and methods have had any considerable place in merchant shipbuilding, does not extend back over very many years. In connection with the Royal Navy, however, the study of scientific naval architecture has been fostered and promoted under Government auspices almost from the commencement of the present century; not, however—it must be added—without alternating periods of regard and neglect, nor irrespective of pressure from extraneous sources.

Although progress in this matter has not been solely due to Government agencies, it may be maintained that a large part of the positive and accurate scientific knowledge which now exists has grown out of the exigencies of the naval service, and has come from sources more or less supported by or connected with Government institutions. It will of course be understood that the science of naval architecture is a field in which many besides shipbuilders, and indeed many besides professional naval architects, have laboured with signal success. The fund of knowledge has been enriched, and the practice of

PORTRAIT
AND
BIOGRAPHICAL NOTE.

———

WILLIAM H. WHITE.

WILLIAM HENRY WHITE.

FELLOW OF THE ROYAL SCHOOL OF NAVAL ARCHITECTURE ; MEMBER
OF THE COUNCIL OF THE INSTITUTION OF NAVAL ARCHITECTS ; MEMBER
OF THE INSTITUTION OF CIVIL ENGINEERS ; AND OF THE
ROYAL UNITED SERVICE INSTITUTION ; LATE CHIEF
CONSTRUCTOR OF THE ROYAL NAVY.

BORN at Devonport in 1845. Entered the Royal Dockyard, Devonport, in 1859. Appointed to an Admiralty Scholarship in the Mathematical School there in 1863, and received a preliminary training in shipbuilding, ship-drawing, and applied mathematics. In 1864 appointed an Admiralty student in the Royal School of Naval Architecture and Marine Engineering, South Kensington, standing first in the competitive entrance examination, and maintaining the first place throughout the course of training. Received his diploma of Fellowship (first class) of the Royal School of Naval Architecture in 1867, and was at once appointed to the Constructive Department of the Admiralty. From 1867 to 1883 continued in the Royal Navy Service, and attached to the Admiralty Department, rising to be Secretary to the Council of Construction in 1873, Assistant Constructor in 1875, and Chief Constructor in 1881. Was appointed Professor of Naval Architecture at the Royal School of Naval Architecture in 1870, and continued to hold that position at South Kensington, and at the Royal Naval College, Greenwich, until 1881, concurrently with his appointment at the Admiralty. Resigned his position in the public service in March, 1883, in order to assume the office of Naval Constructor to the firm of Sir W. G. Armstrong, Mitchell & Co. (Limited), Newcastle-on-Tyne. Is the author of " A Manual of Naval Architecture," well known and highly valued by all classes in the profession, and of numerous papers on professional subjects separately published, or read before the Institution of Naval Architects, the Royal United Service Institution, and kindred Societies.

Yours truly

J W White

shipbuilding improved, by men whose association with the shipyard has been of an indirect and amateur kind, and—it must be added—whose valuable labours the shipyard has often but scantily recognised. . Mathematicians—"mere theorists," as they have been called—have made original investigations and scientific analyses which have upset many previously received practical notions, and established principles, the appreciation of which alone, has led to subsequent progress in actual practice. The part taken by merchant shipbuilders has consisted in the experimental verification, and sometimes the practical correction of principles thus evolved, but even to this extent the service done has been largely incidental. Those considerations which form the economic basis of every commercial concern have naturally circumscribed such service, and only a few notable firms have been able to break through the common restrictions.

The systematic study of scientific naval architecture may be said only to have begun in Britain in 1811, in which year, as the outcome of recommendations made by a Government Commission appointed to inquire into naval construction in 1806, the first School of Naval Architecture was established at Portsmouth, under the direction of Dr. Inman, a distinguished member of the University of Cambridge. All the great advances which had been made previously in the science of naval architecture were chiefly due to foreigners, and any one wishing to acquaint himself at first hand with all that was then most advanced would have to consult the learned treatises of such distinguished Frenchmen as Bouguer, Dupin, Euler, D'Alembert, and the Abbé Bossut, of the distinguished Spaniard Don Juan d'Ulloa, and of Chapman, the celebrated constructor of the Swedish Navy. One or two English writers, between 1750 and 1800, had published translations of some of these foreign treatises, but the only original work of any importance was by Atwood, who contributed a " Disquisition on the Stability of Ships " to the proceedings of the Royal Society (1796-98). This contribution was both a criticism and an extension of flotation and stability investigations by Bouguer,

and as an example of scientific method applied to exact calculations of the qualities of ships it is still well worthy of
study.   In 1791 a "Society for the Improvement of Naval
Architecture" had been formed, the membership being both
numerous and influential, and in 1806 the growing sense of need
for improved scientific methods culminated in the appointment
of the Commission above mentioned, and in the establishment
five years later of the first School of Naval Architecture.   This
institution existed for over twenty years, over forty students
were trained, and the science of naval architecture was greatly
promoted through its agency.   Almost as a body the students
of this school, with their able teacher, deserve the honour of
being regarded as the founders of an English literature of naval
architecture.   Nevertheless, the recognition of Dr. Inman's
services, and his pupils' capabilities as designers, by the naval
authorities was of a cold and disappointing nature.   Ultimately,
however, many of them attained positions wherein their talents
found worthy exercise.

After the abolition of the School of Naval Architecture, under
Dr. Inman, in 1832, no agency for higher education existed
until 1848, when the urgent necessity for a steam re-con-
struction of the Navy forced attention to the want of trained
men, and resulted in the establishment of a second school at
Portsmouth.   The principal of this school was Dr. Woolley, an
eminent graduate of the University of Cambridge.   From 1848
on to the present time, Dr. Woolley has held a prominent place
amongst the promoters of naval science, and the pupils produced by the institution under his directorship have given
in various ways good practical evidence of his capability as
a teacher.   After five or six years of useful work, this second
school was done away with, and a third was established in
London in 1864, after pressure had been brought to bear upon
the Government of the day by the Institution of Naval Architects—an association which was founded in 1860, and which
has since had so flourishing an existence.

The new school was placed for a time under the control of
the Science and Art Department at South Kensington, Dr.

Woolley being Inspector-General, and the late Mr. C. W. Merrifield, F.R.S., Principal. This school, unlike its predecessors,
was not nominally a mere Admiralty establishment, but offered
admission to private naval architects and engineers, and did
not exclude foreigners. It remained in operation at South Kensington until 1873, when the Admiralty decided to establish
the Royal Naval College at Greenwich, and to train their
students of naval architecture and marine engineering there.
Since 1873, therefore, what may be regarded as a continuation
of the third school has been at work at Greenwich, the Admiralty granting facilities for the entry of private and foreign
students, much as was done at South Kensington.

The small extent to which this institution has been taken
advantage of by private students, or by those whose aim is
to equip themselves for service in merchant shipbuilding,
notwithstanding the inducements existing in the shape of
substantial scholarships, has often been subject of comment.
Various reasons have been adduced for this state of matters,
but the true cause would seem to be largely concerned with
the character of the entrance examinations and with the
course of study provided. The subject is well worthy of
consideration, and fuller reference will be made to it further
on when some educational agencies which have been recently
established are under consideration.

At such important junctures in the history of shipbuilding
as the introduction of steam power for propulsion in place of
sails, and the employment of iron in place of wood for the
hulls, precedent and experience lost much of their value under
the new conditions. The association of civil and mechanical
engineers with shipbuilding at these crises was of immense
advantage. Such men as Fairbairn and Brunel, who had
previously gained high reputations in other branches, were
enabled by their scientific skill in designing bridges and other
structures in wrought-iron, to achieve much, and to take the
lead in ship design and construction. "To men of this class,"
says Mr. W. H. White, in the article already alluded to,
"careful preliminary investigation and calculation naturally

formed part of the work of designing ships; 'rule of thumb' was not likely to find favour, even if .it had been applicable, which it was not, under the circumstances.   At first, much was done on imperfect methods, comparatively in the dark; failures were not rare; yet progress was made, and gradually greater precision was attained, in the attempt to design steamers capable of proceeding at certain assigned speeds when laden to a given draught. In fact, the construction of steamers rendered imperative a careful study of the laws of fluid resistance, and of the cognate investigation of the mechanical theory of propulsion—both of which subjects lay practically outside the field of the designers of sailing ships.   The speed of a sailing ship is obviously dependent upon the force and direction of the wind; her designer, therefore, chooses forms and proportions which will enable a good spread of canvas to be carried, on a handy stable vessel.   Questions of resistance to the progress of the ship were therefore subordinated to sail-carrying power and handiness in sailing ships; whereas in steamers designed for a certain speed the question of resistance occupies a primary place, seeing that the engine-power must be proportioned to the resistance.   Consequently, while keeping in view stability, handiness, and structural strength, the designer of a steamer has a more difficult task than the designer of a sailing ship, and the difficulty can only be met if faced intelligently by scientific analysis.   Hence it happened, as was previously remarked, that a more general appreciation of the value of scientific methods accompanied the development of steam navigation and iron shipbuilding in the British mercantile marine."

Another name that must be linked with those already mentioned in connection with the change from wood to iron in shipbuilding, and with the new conditions imposed by the transition from sail to steam, is that of the late Mr. John Scott Russell, already referred to at the beginning of this work.   In the fields of inquiry so largely opened up at the period referred to, Mr. Russell was a most distinguished worker.   His advocacy and adoption in practice of special structural principles, as illustrated not only in the *Great*

Eastern but in other vessels, has influenced subsequent practice incalculably, and by his persevering investigations upon the resistance of vessels, and the "wave-line" theory he advanced, as well as by his inquiry into the characteristics of wave-motion, he laid designers of that period and subsequent investigators under great indebtedness. His contributions to the literature of the profession—notably his *magnum opus*, entitled "Modern System of Naval Architecture"—and the large share he subsequently took in the deliberations of the Institution of Naval Architects, and of other societies concerned with shipbuilding and engineering, enhance that indebtedness and remain as permanent records of his skill and originality.

Approaching the period with which this review is more particularly concerned, reference must now be made to the valuable labours of two eminent men, whose loss the profession has had to mourn within recent years. These are the late Professor Macquorn Rankine and the late Mr William Froude, neither of whom was by profession a naval architect, yet both of whom were led by love of the subject to give their matured experience as civil engineers and mathematical experts to the promotion of knowledge in this domain.

Rankine appears to have become specially interested in the problems connected with ship design, after he became Professor of Civil Engineering at Glasgow University in 1855. Conjointly with Mr. Isaac Watts, late Chief Constructor of the Navy, and formerly a student of the first School of Naval Architecture; Mr. F. K. Barnes, now Surveyor of Dockyards, and Chief Constructor of the Navy, and a distinguished student of the second school ; and the late Mr. J. R. Napier, a member of the famous Clyde shipbuilding firm, Prof. Rankine produced in 1866 " Shipbuilding: Theoretical and Practical." This valuable treatise was edited, and for the most part written, by Prof. Rankine, and provides a complete system of information on all branches of shipbuilding and marine engineering, although subsequent progress in certain departments of naval science has made a new edition desirable. The work is also distinguished for its enunciation of several theories connected with the resistance and propulsion of vessels by Prof. Rankine, which have become the accepted basis of modern practice. Of these the mechanical theory of the action of propellers, and the stream-line theory of resistance, are the best known. His investigations and writings on the latter subject were most ably supplemented and confirmed by Mr. Froude, whose beautifully-contrived model experiments, coupled with his discovery of the law by which such experiments can be made to afford reliable data for the resistance of full-sized vessels, have laid the profession under even a heavier load of indebtedness.

This, however, was not the only work of investigation and experiment with which Mr. Froude actively and inseparably identified himself. Taking up a subject which many authorities before him had studied and written upon with but little success —that of the phenomena of wave motion and the oscillation of ships in a sea-way—he propounded and demonstrated at the Institution of Naval Architects in 1861, after much careful thought and experiment, a theory with respect to it which at that time was entirely new and striking, but which has since been firmly established as the sound one.

At first, authorities in the science of naval architecture, like Moseley and Dr. Woolley, regarded the new theory with suspicion and disapproval; Rankine, on the contrary, warmly supported it, and helped to develop it and to answer various objections urged against the hypothesis on which it was based. For nearly twenty years Mr. Froude steadily pursued the inquiry, adding one mathematical investigation to another, carrying out numerous experiments, and making voyages for the purpose of studying the behaviour of ships. Broadly speaking, it may be said that whereas earlier investigations gave to the naval architect the power of making estimates of the buoyancy and stability of ships floating in smooth water, they gave up as altogether hopeless the attempt to predict the behaviour of ships at sea, or to determine the causes which produce heavy rolling. On the other hand, thanks to Mr. Froude, the designer of a ship now knows what precautions to take in order to promote steadiness and good behaviour at sea.

Although the propositions enunciated by Mr Froude were accepted as laws in a wonderfully short time—considering their startling nature—their influence on practice, and especially the practical application of the methods of comparison by which they had been established, have not even yet been brought to anything like their full issue. The work is being continued upon the lines laid down by Mr Froude, amongst others by men whose closer intimacy with the actual affairs of the shipbuilding yard may be expected to yield results which will be more immediately reflected in actual practice.

Passing allusion has already been made to the founding of the Institution of Naval Architects, but an association which has gathered into its membership so largely of all sections of men concerned with shipbuilding and shipping, and absorbs so much of the knowledge and talent in these domains, must have fuller reference made to it. Regarding its foundation, in 1860, Mr. White, in his article in the *Westminster Review*, says:

"The scheme of the Institution was happily conceived and well executed. Amongst its earliest members were found the trained naval architects of the first and second Schools, the leading private shipbuilders and marine engineers, the principal shipbuilding officers of the Dockyards, men of science specially interested in naval architecture, shipowners, merchants, and others connected with shipping; while a considerable number of sailors from the Royal Navy and Mercantile Marine showed their appreciation of the value of naval science by becoming Associates. The list of names is eminently representative. Sir John Pakington (afterwards Lord Hampton), then only recently retired from the office of First Lord of the Admiralty, was the first President. Many experienced naval officers supported him. There were men like Watts, Read, and Moorsom, who had been pupils of Dr Inman half a century before; others, like Fairbairn, Laird, and Grantham, who had been conversant with iron shipbuilding from its commencement; marine engineering was worthily represented by veterans like Penn, Maudslay, and Lloyd; mathematicians and men of science like Canon Moseley, Dr Woolley, Professor Airy, and Mr Froude appear on the list. Private shipbuilders and naval architects like Scott Russell, Samuda, Napier, and White, joined in the movement, so did the surveying staff of Lloyd's Register. In fact, there was a general appreciation of the endeavour to establish an association which should enable all classes interested in shipping to interchange ideas and experience with a view to general improvement. Mr Reed was the first Secretary, retaining that post until he was appointed Chief Constructor of the Navy, and in that position did much to aid the progress of the Institution."

While it is true that the membership list of the Institution in its early days was of the representative character above indicated, it should be pointed out that the actual proceedings of the Institution were not shared in by anything like the variety of talent which the list comprised, or which now distinguishes its annual meetings. For many years it was almost the exclusive conference of Admiralty authorities and members of those shipbuilding and engineering firms who undertook Government work, and the transactions for a long time were very largely confined to purely naval matters. The scientific value of the earlier volumes of the transactions would certainly have suffered considerably if the papers by Mr. Froude and Prof. Rankine had not formed contributions, and the prosperity and development of the Institution would have been equally lessened had there not been general infusion of " new blood" from the mercantile marine in all parts of the country. This has been going on during the past twelve years or more, and the scope and utility of the Institution's proceedings have increased with the change. Of the later development of the Institution, the authority already quoted says:—

"Owing to the rapid advances constantly being made in both the science and the practice of the profession, the 'Transactions' have come to be the chief text-books available. Members and Associates have joined from all the great maritime nations. Members of the professional corps of naval architects and engineers of France, Austria, Italy; Germany, the United States, Russia, Sweden, Norway, Denmark, Holland, are proud to be numbered with their English professional brethren, and not a few of these foreign members have contributed valuable Papers. The meetings of the Institution afford exceptional opportunities for the discussion of questions having general interest, as well as others having more special value to professional men. Different views of the same subject find capable exponents, and lead to valuable discussions. The latest systems of construction and most recent changes in *materiel* are described by competent authorities. Valuable *data* are put on record relating to the designs and performances of war-ships and merchant-ships. Inventions of various kinds are described and examined. Abstruse theoretical investigations are by no means rare; and, in many cases, the contribution of one such Paper by an original thinker has given a start to others and led to important extensions of knowledge. In fact, the Institution of Naval Architects has admirably fulfilled the intentions of its founders, acting as a centre where valuable information could be collected, and whence it could be distributed for the general benefit of the profession. Before it was founded naval science had no home in England; its treasures lay scattered far and wide in occasional Memoirs and Papers; but now everything worth preservation naturally finds its way to the 'Transactions.' Any movement affecting shipping also leaves its record there in Papers and Discussions which will hereafter have a high historical value."

As evidencing the change which has latterly come over the Institution with respect to its annual proceedings, it may be noted that whereas in the early years there were at some meetings no papers—leaving out of account, those by Froude and Rankine—except by Admiralty members and others concerned with Government work, there was not a single paper by an Admiralty man during the meetings of the present year.

With the general reference already made to Mr Froude's invaluable labours in connection with the resistance of vessels the brief statement of the agencies through which progress has been made during the present century may be considered as brought down to the period coming within the scope of the term "Modern," as used in this work. The more difficult task of chronicling the progress made during the period in question, both in the science of naval architecture purely, and in the application of science to practice, must now be attempted. The plan upon which it is proposed to accomplish this is to show wherein and to what extent scientific methods in designing and observing the behaviour of ships have been regarded, and indicating generally where still further improvement may be looked for. To accomplish this in such a way as to take appreciative account of the most salient features, and yet to avoid difficult technical terms and unnecessary elaboration, may involve some omissions and slight inaccuracies, important enough from a strictly scientific point of view, yet which do not materially affect the faithfulness of the record.*

As preparing the way for references to those more special points in connection with which scientific progress has taken place during recent years, the following general and ele-

---

* In this, as in other matters dealt with, the full appreciation of which involves careful technical study, readers are referred to the papers enumerated at the end of chapter, as well as to the "Manuals" already referred to in this work.

mentary outlines of the principal scientific problems in ship design and construction may be helpful to many readers:—

DISPLACEMENT AND CARRYING CAPABILITY.

A vessel floating at rest displaces a volume of water whose weight equals her own total weight.

For vessels floating in sea-water the number of cubic feet of water displaced per ton of weight is, as nearly as possible, thirty-five. For vessels in fresh water—*i.e.*, lakes or rivers—the cubic feet per ton of weight is thirty-six.

By calculating the volume of under-water portion of the vessel's hull, the number of cubic feet displaced by the vessel when floating at any given draught is obtained. This result, divided by 35 or 36, according as the water is salt or fresh, gives the number of tons weight displaced, and consequently the total weight of the vessel.

Calculations being made of the volume of the vessel's hull to intermediate distances between the keel and the maximum load-line, it is thus possible to construct a "curve of displacement" from which the actual amount of displacement at any intermediate draught can be obtained.

From this curve a set of scales—usually set up alongside a vertical scale of feet and inches, representing the vessel's draught-marks—are constructed, showing—1st, the tons "displacement" at any draught; 2nd, the tons of "deadweight" capability—*i.e.,* the tons displacement due to the weight of cargo, coal, ballast, stores, fresh water, spare gear, &c.—at any draught above the vessel's light draught : "light draught" being that at which the vessel floats with holds clean-swept, bilges dry, water in boilers, and with such spare gear on board as is required by Board of Trade; and 3rd, the amount of "freeboard"—*i.e.*, the distance in feet and inches from any particular draught line to the top of the deck amidships.

BUOYANCY AND STABILITY.

A ship floating upright and at rest in still water must fulfil two conditions—1st, as stated above, she must displace a weight of water equal to her own weight; 2nd, her centre of gravity must lie in the same vertical line with the centre of gravity of the volume of displacement or "centre of buoyancy."

The whole weight of the ship may be supposed to be concentrated at her centre of gravity, and to act vertically downwards, and the

resultant vertical pressure of the surrounding water in the same
way to act upwards through the centre of buoyancy.

. When the ship has been inclined from the upright position, by any
force, the downward and the upward forces—weight and buoyancy
respectively—act through two separate but parallel vertical lines,
and form what is technically known as a "couple." The perpendicular distance between the vertical lines usually varies with the
inclination, and is called the "arm" of the couple. This arm
measures the leverage with which the weight and buoyancy of the
ship tend either to force her back into the upright position, or to
incline her still further, and, it may be, to capsize her. The former
effect would be the result of what is known as a "righting couple,"
the latter the result of an "upsetting couple."

FIG. 14.   **FIG. 15.**

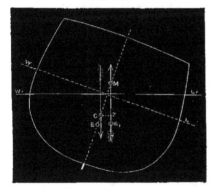 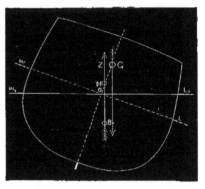

This may be made clearer by illustration. On Figs. 14 and 15,
which show in outline a vessel's midship section, the vessel being
inclined to a small angle, G represents the centre of gravity of
vessel, and B the centre of buoyancy. The water line W.L. corresponding to the upright position, in the inclined position becomes
W ₁. L ₁., and the centre of buoyancy B shifts out on the immersed
side of the vessel to B ₁. Assuming in the case of Fig. 14 that
some external force not involving any shifting of the centre of
gravity has produced the inclination, then the weight of the vessel
acts downwards through G, and the buoyancy of her displacement
acts upwards through B ₁., as indicated by the arrows passing through
these points. The combined effect of these forces, in this case, is
to rotate the vessel towards the upright, *i.e.*, it forms a "righting
couple." Fig. 15 illustrates a case of the opposite kind. The angle
of inclination may be supposed to be greater than in Fig. 14, and

the centre of gravity G is much higher in the vessel. The vertical through B ı is to the left instead of to the right of the vertical through G. The effect of the forces in this case is to rotate the vessel in the direction of inclining her still further, and to capsize her—*i.e.*, it forms an "upsetting couple." A line at G, therefore (Fig. 14), taken at right angles to the new vertical line, gives the distance which corresponds to the righting arm (G Z). A similar line at G (Fig. 15) represents the upsetting arm. The lengths of these arms when multiplied into the displacement, gives the "moments" at the respective degrees of inclination. The "curve of stability" for a vessel is simply a graphic representation of these arms or moments. When calculated for the various degrees of inclination, they are set off as ordinates along a base line—the righting arms or moments above, and the upsetting arms or moments below, the line —at distances corresponding to the number of degrees in the respective inclinations. A curve drawn through the extremities of these ordinates is the curve of stability.

The two points above named whose relative positions are vitally concerned with this subject—*i.e.*, centre of buoyancy and centre of gravity—are determined by shipbuilders for many of their vessels, although the stability may not be calculated to its full extent. The position of the centre of buoyancy is easily ascertained from, and in fact usually forms part of, the displacement calculation. While the position of centre of gravity may be found by means of calculation alone, *i.e.*—by the process of estimating the position of the centre of gravity of each of the component parts, and from this deducing the common centre of gravity of the whole ship—the work is so laborious, complex, and so liable to error, that it is scarcely ever adopted at the present day by mercantile shipbuilders. The position can be ascertained with comparative ease and greater accuracy by means of "inclining" experiments with the finished vessel, or closely estimated before-hand by means of data obtained in the manner alluded to from previous vessels of similar type.*

Another point concerned with stability is that termed the "meta-

---

* The principle which underlies the experiment is this :—If any one body forming part of a system of heavy bodies be moved from one position in the system to another, the weight of the body moved multiplied into the distance through which it is moved, is precisely equal to the weight of the whole system of bodies multiplied into the distance through which the

centre," which is found by calculation from the lines of the vessel. Referring to Fig. 14, a vertical line .drawn through the centre of buoyancy B₁ cuts the original vertical line at **M**. The intersection **M**, when the vessel is inclined to an indefinitely small angle, is the "metacentre." It is approximately the same in all ordinary vessels for inclinations less than say 10°, but varies with greater inclinations. The corresponding intersections of the consecutive vertical lines for all degrees of inclination are embraced in the term "metacentrique." These features in stability investigations were originated by Bouguer, to whom reference has already been made. The manner in which they are concerned with stability will be indicated further on. (See also footnote on preceding page.)

RESISTANCE POWER AND SPEED.

A ship, in moving through the water, experiences resistance due to a combination of causes, which combination, according to modern accepted theory, is made up of three principal elements.

1st—"Frictional" or "skin-friction" resistance, due to the particles of water rubbing against the ship's hull ;

2nd—"Eddy-making" resistance, due to local disturbances or eddies amongst the particles of water—almost wholly at stern of ship.

3rd—Surface disturbance of the water by the passage of the ship, resulting in the creation and maintenance of waves : known as "wave-making" resistance.

---

common centre of gravity of the whole has moved. If in a ship, therefore, a movable weight of known amount is moved across the deck through a given known distance, the centre of gravity of the ship itself, with all on board, has been moved in a line parallel to that through which the small weight has been transferred, and through a distance inversely proportioned to the weight of the whole ship to the weight moved. If, for instance, a weight of five tons should be moved through a distance of twenty feet, then multiplying this weight into this distance and dividing by the total weight of the ship, the distance through which the ship's centre of gravity has travelled parallel to the deck is obtained. If, at the same time, an exact measure of the angle through which the ship has been inclined by moving the five tons through the distance named has been taken, and the position of the ship's metacentre has been obtained, then the elements of a triangle are known—namely, the degrees in each of its angles, and the length of one of the sides—and from these the length of the remaining sides of the triangle is easily deduced. One of these sides will be the distance between the metacentre of the ship and its centre of gravity, and, consequently, the metacentre being known from calculation, the position of the centre of gravity becomes known also.

The conditions which govern each of these elements, and their relative importance, may be generally indicated.

Surface-friction resistance, especially for vessels moving at moderate or slow speeds, is much greater than the resistance due to other causes—that is if the hull is ordinarily well formed. Its amount depends upon the area of the immersed surface, upon its length, upon its degree of roughness, and upon the velocity with which the water glides over it—*i.e.*, upon the speed of the vessel.

Eddy-making resistance only acquires importance in exceptional cases, *e.g.*, in ships having unusually full sterns. In ordinary well-formed ships it is of small amount, and is caused mainly by blunt projections such as shaft tubes, propeller brackets, and stern posts.

Wave-making resistance is much more variable than surface-friction resistance. Its amount depends on the form and proportions of vessels, and on the speed at which they move : being greatest, of course, in ships of full form and in those moving at high speeds.

The sum of these three main elements of resistance constitutes the total resistance experienced by a vessel if "towed" through the water, that is, the resistance considered apart from the action or influence of the propelling instrument. In the case of a steamship, however, propelled by a screw or paddle wheels, the resistance is augmented, more or less considerably, according to the form, surface, and disposition of the propelling instrument.

By the employment of various formulæ deduced by scientific authorities from theory and experiment, an approximation can be made before-hand to the total resistance of a proposed vessel, and from this an estimate of the power required to drive her at a certain speed. Moreover, through the law of comparison propounded by Mr Froude, the resistance of a ship can at all times be deduced with fair accuracy from the resistance of her model, certain corrections well determined by experiment having to be made.

The power of marine engines is expressed either in "nominal" or "indicated" horse-power. Nominal horse-power is a term practically obsolete so far as being a measure of the efficiency of engines, and only exists as a conventional method of commercially measuring the sizes of engines. Indicated horse-power measures the work done by the steam in the cylinders during a unit of time, and 33,000 units of work per minute, or 550 units of work per second, constitute one horse-power. The effective mean pressure of the steam is ascertained from diagrams drawn by means of the instrument known as the "Steam Engine Indicator," and hence the term "indicated" horse-power.

The development by a vessel's engines of the power requisite to drive her at a certain speed is always very considerably more than the power required simply to overcome her total resistance at that speed. This excess of power developed over power usefully employed in overcoming resistance is known as "waste work." It amounts in many cases to as much as from 50 to 60 per cent. of the gross indicated power, and it is absorbed mainly as follows :—In overcoming frictional and other resistances of the engines and shafting, working air pumps, &c., and in overcoming the frictional and edgeways resistance of the propeller. The residue of power usefully employed is known as the 'effective' horse power. The respective causes of ' waste ' and their relative amounts are problems constantly demanding solution. Progressive speed trials with actual vessels and experiments with small scale models are daily contributing to their solution, and to some extent to their reduction.

STRUCTURAL STRENGTH.

Considering a ship as floating in a state of rest in still water, the volume of displacement represents a weight of water equal to the weight of the ship. This equality, however, does not exist evenly throughout the length of the vessel, or for individual portions: thus, amidships the weight of water displaced by a given length —in other words, the buoyancy—is usually considerably in excess of the weight of that portion of the vessel and her contents. Similarly at the extremities the 'weight' of a certain length exceeds the 'buoyancy.' Between the part or parts of the vessel in which there is excess of buoyancy over weight, and the part or parts in which the weight exceeds the buoyancy, there must obviously be sections of the ship at which the two are equal, and these are termed "water borne" sections. A ship circumstanced as described is in a condition similar to that of a beam supported at the middle and loaded at each end. Such a beam tends to become curved, the ends dropping relatively to the middle, and the ends of the ship tend to drop similarly, the change of form being called "hogging." On the other hand, if the excess of buoyancy occurred at the extremities and that of weight amidship, the ship would resemble a beam supported at the ends and loaded at the middle. In such a condition the middle would tend to drop relatively to the ends: a change of form called "sagging."

These general principles are much more readily and safely applicable to ships while floating in 'still water' than to ships when at sea—the strains experienced then being necessarily the results of far more complex and severe influences. The existence of waves and their rapid motions relatively to that of the vessel, and the pitching, heaving, and other movements thus caused, increase the inequality of distribution of weight and buoyancy and affect more materially the strains brought upon vessels. Consideration of the problem, therefore, involves a study of waves, both as to their formation and action, and necessarily leads to a mode of treatment which cannot have accurate regard for particular cases. Variable influences of immense importance are also constituted by the state of loading in vessels for merchant service. For a uniform basis of comparison in these calculations such vessels are usually assumed as loaded with homogeneous cargo—*i.e.*, cargoes of equal density throughout.

This fundamental element of relative 'weight' and 'buoyancy' having been indicated, the chief strains to which a ship is subjected may now be stated. This may be done with sufficient regard to general accuracy, under four heads:—*

(1) Strains tending to produce longitudinal bending—"hogging" or "sagging"—in the structure considered as a whole.

(2) Strains tending to alter the transverse form of a ship, *i.e.*, to change the form of athwartship sections.

(3) Strains incidental to propulsion by steam or sails.

(4) Strains affecting particular parts of a ship, or "local strains" —tending to produce local damage or change of form independently of changes in the structure considered as a whole.

To these might be added various other strains, which, however, are of less practical importance, and are not felt in any great degree —except in very special cases and under unusual circumstances— apart from the strains which affect the structure considered as a whole. The provisions made for the latter are, under ordinary circumstances, sufficient to cover the demands of the former, but particular cases may have to be provided for on their merits, apart from the treatment generally applicable.

---

* The classification of strains here given is as contained in White's "Manual of Naval Architecture." To this authoritative source readers must turn who wish a full exposition of the several problems so shortly dealt with in these pages.

The manner of ascertaining the strength of a ship to resist strains tending to produce longitudinal bending, is to compute the effective sectional area of all the longitudinal items in the structure which are brought under compressive or tensile strain, and from this to calculate the strength in the same manner as for a girder having an aggregate sectional area and a disposition of material equivalent to that of the ship.

To ascertain the accurate maximum strains tending to produce longitudinal bending, or, in excessive cases, to break the ship across at the transverse section where the strains reach their maximum, involves a careful and most laborious consideration of the relative weight and buoyancy of individual sections throughout the length, and is a task not generally undertaken in mercantile shipyards.*

References to the nature of the transverse and other strains above enumerated and the extent to which they have been investigated will be made further on.

With regard to such fundamental properties of vessels as displacement, weight, and carrying capability, nothing new has for a long period been added to the fund of scientific knowledge. One of the conditions now most commonly laid down by the owners of a proposed ship is that which provides for a certain carrying capability on a given draught of water and at a certain speed, the principal dimensions of the vessel also being stipulated. The problem of determining what total displacement will be required, involves consideration and an estimate of—1st, The total weight of hull having regard to structural strength; 2nd, the total weight of machinery having regard to speed required. By using "co-efficients" deduced from the weights of vessels of similar type already built,† these are

---

* This will be more fully referred to further on, but it may be stated here that the need for independent calculation is largely obviated, owing to the existence of "co-efficients," deduced from investigations made by experts. Further, the existence and influence of the Registration Societies are such that the codes of scantling and the structural supervision instituted by them together constitute the only guarantee of structural strength generally desiderated.

† Suppose the dimensions of a proposed vessel to be 320 × 36 × 26½ feet, then, according to a method of approximation largely in use, the sum of these dimensions divided by 100 gives what is known as the "cubic number"—(320

determined; and adding them to the carrying capability or
deadweight stipulated, the required displacement can be closely
approximated to.   For vessels of abnormal proportions or of
very unusual construction careful and detailed calculations of
the weight of materials are undertaken previous to tendering
for them.   In some yards, indeed, a like degree of care is
observed in ordinary cases: methods of approximation involving the use of co-efficients such as that based on cubic capacity
being distrusted.

The further problem of determining what form of hull will
give the required displacement is the essential and all-embrac-
ing feature of the work of design, as it involves consideration
of almost all other properties.   The methods of designing ships
are various, and a very common method, at one time more
followed than it now is, consists in shaping a block model
direct, and'from it taking the necessary measurements for displacement, and for full-size delineation in the moulding-loft.
The disadvantages pertaining to this somewhat antiquated
method are becoming more recognised as shortened and exact
methods of linear or "draught plan" design are put forward.

Unless the plan of lines of a similar vessel of nearly the
same dimensions is at hand, the design of a new vessel is
in many instances done without previous calculation being
made to ensure at once obtaining the desired displacement.
Special methods of quickly arriving at this result are, however,
not uncommon in mercantile shipyards, and generally speaking
the chief draughtsmen in the employ of large firms doing a
varied class of work have rules derived from long experience,
though not perhaps definitely systematised, by which they are

---

× 36 × 26½ ÷ 100) = 3052 cubic number.   Suppose that for a vessel already
built, similar in type and dimensions, or of similar proportions, to the one
proposed, the cubic number, when divided into the ship's actual weight
(*i.e.*, the displacement *minus* the weight of machinery and the deadweight
carried), gives say ·53, then this figure represents the "co-efficient" of ship's
weight, and applying it in the case supposed gives:—3052 × ·53 = 1620, the
weight of hull for proposed vessel.   This example illustrates the manner in
which the weight of machinery is estimated, and indicates the nature and use
of the general term "co-efficient:" frequently employed in this chapter.

guided.* Irrespective of all such special methods, however, the work of designing is now greatly shortened and simplified by means of Amsler's "planimeter," an ingenious instrument for measuring areas now becoming well known. † By employing the instrument in question, the draughtsman need not too laboriously strive after the exact displacement at first, as the time occupied in ascertaining what displacement any set of lines gives, and in the consequent fining or filling out, is very considerably less than by the ordinary methods.

The question of stability, which has next to be considered, is one of great difficulty and intricacy, and it was not till the middle of last century that some of the principles upon which it depends began to be understood. Bouguer showed in 1746 that the position of the "metacentre" limits the height to which the centre of gravity of a floating body may be raised without making it unstable, and that the righting moments at small angles of inclination from a position of stable equilibrium are proportional to the height of the metacentre above the centre of gravity. As the position of the metacentre for any given draught of water is easily determinable when once the volume of displacement and the centre of buoyancy at that draught

---

* One such method, devised and followed by Mr. C Zimmermann in his daily practice as chief draughtsman to the Barrow Shipbuilding Company, and described by him before the Institution of Naval Architects in 1883, gives with very little preliminary calculation, and at once, a close approximation to the correct displacement. Another system, originated and used in practice by Mr. Chas. H. Johnson, chief designer to Messrs Wm. Denny & Brothers, consists of an analysis of the lines of vessels of various degrees of fineness and fulness previously built, formulated for daily use in a series of curves of areas, giving, for sections at certain fixed distances from midships—in terms of percentage to the midship area—the particular area specially suited to afford the required displacement; and at the same time to maintain the general form of hull which in actual practice has proved satisfactory with respect to speed. In his later practice, Mr. Johnson has found it preferable to use the block form of analysis of Mr. A. C. Kirk (considered further on in matters relating to speed), using the three sides of that form as a basis upon which to group the waterlines.

† For illustrated descriptions of this and other improved calculating instruments referred to in this chapter, see Appendix.

have been ascertained, it has been the practice for a very long time to construct a curve representing the height of the metacentre at all draughts, and to use it for showing the limits above which the centre of gravity cannot be raised with due regard to the stability required for the practical working of vessels and for purposes of safety. By the method of "inclining" vessels, already described (see outline of fundamental principles, page 98), the determination of the precise position of the centre of gravity is rendered comparatively simple.*

While the vertical distance between the centre of gravity and the metacentre—commonly termed the "metacentric height" —forms a measure of the " initial stability," or the stability at very small angles of inclination, it is imperfect by itself, and may be very misleading as regards the stability at larger angles. This was conclusively demonstrated by Atwood in his papers read before the Royal Society in 1796 and 1798, while other grounds for discrediting the standard of stability furnished by mere metacentric height were discovered subsequently, and have been signally emphasised, with additional reasons, by recent occurrences. Atwood, in the papers referred to, laid down a general theorem for determining the righting moments at any required angles of inclination possessed by a ship having a given draught of water and a fixed height of centre of gravity, the principle of which involved the use of the moments of the volumes of the " Wedges," *i.e.*, those parts of a vessel (see W O Wi, Li O L, fig. 15), which become immersed and emerged as

---

* This experimental method, it may be explained, has long been practised in connection with ships built for the Royal Navy, and for a considerable number of years it has been systematically followed in some leading merchant shipyards. Messrs A. & J. Inglis, Pointhouse, Glasgow, and Messrs Wm. Denny & Bros., Dumbarton, were amongst the earliest firms to systematically adopt the practice. With the former it has been customary to incline every vessel of distinctive type built by them since 1871, and with the latter the practice has been constantly followed from a date somewhat subsequent. For some years past other firms on the Clyde and elsewhere have adopted the method, the data so accumulated being found an admirable basis from which to estimate the height of the centre of gravity in proposed vessels. Tables giving the results of inclining experiments made on various types of merchant steamships and sai'ing vessels will be found in "White's Manual of Naval Architecture," pages 82-87.

PORTRAIT
AND
BIOGRAPHICAL NOTE

———

JOHN INGLIS, Jun.

JOHN INGLIS, Jun.,

MEMBER OF COUNCIL OF THE INSTITUTION OF NAVAL
ARCHITECTS; MEMBER OF THE INSTITUTION OF ENGINEERS AND
SHIPBUILDERS IN SCOTLAND, ETC.

BORN in Glasgow in 1842, where his father, Mr Anthony Inglis, and Mr John Inglis, his uncle, were marine engineers, subsequently also becoming iron shipbuilders. Under the designation of A. & J. Inglis the combined businesses — the engineering works at Warroch Street, and the shipyard at Pointhouse — have been conducted with marked success. Having for some years attended the Glasgow Academy, Mr INGLIS, at the age of fifteen, entered the University, where for several sessions he studied under such teachers as the late Professors Ramsay, Blackburn, and Rankine, and also under Sir William Thomson. Of Professor Blackburn's mathematical and Professor Rankine's engineering classes Mr INGLIS was a distinguished student; in the former—although the youngest on the roll—carrying off several prizes, and in the latter acquiring a sound knowledge of applied mathematics as concerned with engineering and naval architecture. This experience was afterwards supplemented by a term's apprenticeship in the practical work of the engine shop. The art of naval construction, however, had always irresistible attraction for Mr INGLIS, and in 1867 he seriously applied himself to the concerns of the shipyard, taking an active share in its management ever since. Mr INGLIS' career, though uneventful, has been one of assiduous devotion to the profession of Naval Architecture, especially as directed to scientific investigation and analysis. The fruits of this are reflected in many noteworthy and specialized steam vessels produced by his firm. Was the first shipbuilder on the Clyde to follow the practice of inclining vessels to ascertain their stability, and was one of the earliest on the Clyde to apply the correct method of estimating longitudinal strains to the hulls of steamers. His firm have been noted for the careful and elaborate trials of steamers on the measured mile, and the digesting of such data. Is the author of several papers read before the societies with which he is connected, one of which fully described the system of speed trial and analysis above referred to. The designing and sailing of yachts are favourite pursuits of Mr INGLIS; and the system of yacht ballasting by means of a lead keel forming portion of the hull structure was first instituted by him in one of the many yachts built for his own use. Under the title of "A Yachtsman's Holidays," he published, some years ago, a volume giving a racy account of yachting experiences in the West Hebrides. He wields a forcible pen, and it is not unfrequently employed anonymously in the interests of shipbuilding and naval science.

Yours faithfully
John Inglis Junr

she is inclined. Several methods of simplifying Atwood's calculations had been devised previous to 1861,* but in that year Mr. F. K. Barnes, in a paper read before the Institution of Naval Architects, described a method of accomplishing this which until within recent years has been the one ordinarily adopted in computing the stability of a vessel at various angles of inclination.†

Owing to questions having arisen at the Admiralty in 1867 respecting the stability of some low freeboard monitors at very large angles of inclination, Sir E. J. Reed, then Chief Constructor, directed the matter to be investigated. The work was placed in the hands of Mr. William John, who embodied for the first time the results of the calculations in the form of a curve of stability, which exhibited the variations of righting moments with angles of inclination up to the particular angle at which stability vanished. The entire range of a vessel's stability was thus made evident, and in such a form as enabled the general problem to be far more comprehensively and accurately treated than before. The results of Mr. John's labours were described in a paper read by Sir E. J. Reed before the Institution of Naval Architects in 1868, and a further paper, containing an improved method of applying Atwood's theorem to the calculation of stability upon this extended scale, was read before the same Institution by Messrs John and W. H. White in 1871. The loss of H.M.S. *Captain*, in

---

* From the first volume (1860) of the Transactions of the Institution of Naval Architects, it is seen that Dr. Inman, Samuel Read, and Dr. Woolley had each already found different methods of simplifying Atwood's calculations.

† Various other methods of simplifying the calculations based on Atwood's theorem were subsequently proposed, and one or two different methods also brought forward—notably one in 1876 by the late Mr. Charles W. Merrifield, afterwards improved by the late Professor Rankine, and one by Mr J. Macfarlane Gray, of the Board of Trade, described by that gentleman in 1875, but since considerably improved. Most of them were laid before the Institution of Naval Architects in papers which will be found enumerated in the list at end of chapter. While such propositions did not contribute directly to bring the problem of stability to its presently accepted form, they deserve to be remembered as tokens of the great labour and skill which have been expended in founding and developing this branch of scientific naval architecture.

1870, as already pointed out near the beginning of this chapter, occasioned an immediate and serious regard for the stability of war vessels. This disaster, with other losses at sea from instability, also forcibly directed the attention of mercantile naval architects to the subject, and investigations on the same complete scale as those undertaken in the Admiralty have for some years been adopted in a few leading mercantile shipyards.

In this way the peculiar dangers attaching to low freeboard, especially when associated with a high centre of gravity, have been pretty fully made known, but the character of the stability which is often to be found associated with very light draught appears to have escaped the attention it demands. Light draught is often as unfavourable to stability as low freeboard, and in some cases more so.

These truths were forced into prominence at the inquiry held by Sir E. J. Reed on behalf of the Government into the disaster which befell the *Daphne*, a screw steamer of 460 tons gross register, which capsized in the middle of the Clyde immediately on being launched from the yard of the builders, Messrs Alexander Stephen & Sons, Linthouse, on July 3rd, 1883. Sir E. J. Reed, in his exhaustive report, published in August, 1883, emphasised the lessons adduced at the inquiry as to the peculiar dangers attaching to light-draught stability; and Mr. Francis Elgar, (now Professor of Naval Architecture in Glasgow University), who was employed to make investigations respecting the stability possessed by the *Daphne* at the time of the disaster, did much to guide consideration of the subject into this channel. In a letter to the *Times* on 1st September, 1883, Mr. Elgar, by way of explaining portions of his evidence at the inquiry, called attention to the relation which exists between the righting moments at deep and light draughts in certain elementary forms of floating bodies, his communication throwing further light on the subject of light draught stability. It appears that the fundamental proposition which underlies the variations in the stability of a floating body with draught of water had never before been demonstrated or enunciated.

It will be readily understood that a curve of stability for a

PORTRAIT
AND
BIOGRAPHICAL NOTE.

———

SIR EDWARD J. REED.

SIR EDWARD J. REED, K.C.B., F.R.S., M.P.

VICE-PRESIDENT OF THE INSTITUTION OF NAVAL ARCHITECTS ; MEMBER
OF COUNCIL OF THE INSTITUTION OF CIVIL ENGINEERS, AND MEMBER
OF THE INSTITUTION OF MECHANICAL ENGINEERS.

BORN at Sheerness, September 20th, 1830. Educated at the School of Mathematics and Naval Construction, Portsmouth, and served in the Royal Dockyard, Sheerness. Leaving the Government service, he became the editor of the "Mechanics' Magazine," in which position he first became known as an authority on Naval Architecture. Was one of the originators of the Institution of Naval Architects in 1860, and for a number of years acted as Secretary to that body. Submitted proposals to the Admiralty concerning the construction of iron-clad ships, which were adopted in practice, and were so highly approved by the Board of Admiralty that their author was appointed Chief Constructor of the Royal Navy in 1863. During the time he held that office, designed iron-clad ships and vessels of war of every class for the British Navy, and also—with the consent of the Government—some iron-clad frigates for the Turkish Navy. In consequence of his objections to rigged sea-going turret ships with low freeboard, of the "Captain" class, and of the favour that type of ship found with the Board of Admiralty, resigned his office in July, 1870—a step rendered remarkably significant by the lamentable capsizing of the "Captain" two months later. Since his resignation, has designed iron-clad vessels and other classes of war ships for various Foreign Powers ; numerous steam yachts, and smaller vessels. Has recently devised and patented a method of construction for war ships which will reduce to a minimum the destructive effect of marine torpedoes, and which promises to revolutionise present structural systems. Is the author of "Shipbuilding in Iron and Steel," "Our Iron-clad Ships," "Our Naval Coast Defences," "Japan: Its History, Traditions, and Religions," as well as of several papers contributed to the Institutions with which he is connected. Since his retirement from the Admiralty has received numerous recognitions of his professional skill and ability, including various decorations from Foreign Powers. Was created a Knight Commander of the Bath, in 1880. In 1874 was returned to Parliament in the Liberal interest as Member for the Pembroke Boroughs, which he represented till 1880, when he was elected for the important constituency of Cardiff. During the summer of 1883 was deputed by the Government to investigate and report upon the "Daphne" catastrophe on the Clyde, the results of which are elsewhere referred to in this work. In February of the present year was entrusted with the Presidency of the Committee appointed to enquire into the subject of the Load-Line of vessels.

Yours truly,

E. J. Reed.

given draught of water and position of centre of gravity ceases to be applicable if changes are made in the weight and consequent draught of water of a ship or the position of the centre of gravity, or in both. Now in mercantile steamers, from the extremely light condition in which they are launched to the uncertain loaded condition of their daily service as cargo-carriers, the variation of draught is very considerable, and imports into the subject considerations which do not obtain to any great extent in war ships.

To complete the representation of stability as it should be known for merchant ships, it is now recognised that curves showing the stability at every possible draught of water and for different positions of centre of gravity should be constructed. By means of "cross curves" of stability, or curves representing the variation of righting moment, with draught of water at fixed angles of inclination, this comprehensive want can be met with something like the necessary expedition. From such curves it is a simple operation, involving no calculation save measurement, to construct curves of the ordinary description, showing the righting moment at all angles for any fixed draught of water and position of centre of gravity. Professor Elgar was the first to publicly direct attention to this valuable development of stability investigation of merchant ships, doing so in an able paper "On the Variation of Stability with Draught of Water in Ships," read before the Royal Society on March 13th of the present year. Simultaneously with Prof. Elgar's employment of such curves in actual practice their use had been independently instituted by Mr. William Denny in his firm's drawing office, and the mode in which they were worked out in this case was communicated in a paper read by Mr. Denny in April of the present year before the Institution of Naval Architects.* Several important improvements with respect to simplifying and shortening calculation distinguish the method employed by Mr. Denny, and that gentleman, in the paper

---

* "On Cross Curves of Stability; their Uses, and a Method of Constructing Them, Obviating the Necessity for the Usual Correction for the Differences of the Wedges of Immersion and Emersion."

referred to, accords individual credit to members of the scientific staff in his firm's employ, who, on being entrusted with the work of calculation, brought considerable originality to bear upon their labours. The cross-curves described by Prof. Elgar were constructed from a series of curves of stability calculated in the ordinary way. This, however (as pointed out in an after-note to that gentleman's Royal Society paper), is less simple and very much less expeditious than the method carried out under Mr Denny, which consists in calculating the cross-curves directly by applying Amsler's mechanical integrator* to the under-water portion of the ship instead of to the wedges of immersion and emersion, thus determining at once the positions of the vertical lines through the centres of buoyancy at the required angles of inclination. As thus carried out a complete set of cross-curves can be produced with about one-third the labour involved in employing the older method. The ease and rapidity with which ordinary curves for separate draughts can be taken from cross-curves has already been commented upon.

Many other investigators besides those already mentioned have recently been working at the subject of stability, and a considerable number have read papers, dealing with the extension and simplification of stability calculations, before one or other of the scientific societies concerned with naval architecture, most of the methods put forward being well worthy of study.† To very many shipbuilders, however, and to others besides them responsible for the stability of ships, processes of arithmetical calculation—even allowing for all the simplification which mathematical skill has recently effected —appear still to be too intricate, or to absorb too much time for their being entirely followed. As a simple means of readily, although approximately, arriving at the results attained more elaborately and reliably by calculation, attention has recently been directed to an experimental process by which a complete curve of stability may be constructed almost without the use

---

*A detailed description of this valuable instrument will be found in Appendix.

† Space forbids any detailed reference to these, but the names of the papers and their respective authors will be found enumerated in the list at end of chapter.

of a single figure! The method was first brought forward in 1873 by Capt. H. A. Blom, chief constructor of the Norwegian Navy, formerly a student of the South Kensington School of Naval Architecture, who described it to the United Service Institution. The method has been employed by shipbuilding firms on the Tyne and Clyde when a curve of stability had to be produced in a very limited time, and when extreme accuracy was not a desideratum. As practised by the firms in question, the *modus operandi* differs in some slight respects from that described by Captain Blom, but the changes in no way affect the principles as first laid down by him. The modern mode of procedure may be briefly described:—

From the body plan of the ship, *i.e.*, that portion of the draught plan representing the vessel's form by a series of equidistant transverse sections—any convenient number of sections up to the load water-line are pricked upon and then cut out of a sheet of drawing paper of uniform thickness. These sections are then gummed together in their correct relative positions, care being taken to spread the gum thinly and evenly. This paper model—greatly foreshortened, of course—represents the immersed portion of the ship (in other words, the displacement) when she is floating upright. By suspending this model from two different points, and taking the intersection of two vertical lines through the points of suspension—or better still, by balancing it horizontally on a pin and fixing the point when the model is in equilibrium—the centre of gravity of the model, or in other words, the actual centre of buoyancy is obtained.

Water lines at various angles of inclination are then drawn on the body plan, all intersecting the water line for the upright condition at the centre line of ship. The displacement represented by the inclined water lines thus drawn, generally not being equal to that for the upright position, a correcting layer has to be added or subtracted for each inclination, in order to obtain this end. By employing the planimeter the necessary thickness of this layer can be most readily ascertained. Where a planimeter is not available the actual floating line may be obtained, after the model has been made, by cutting off layers, allowance having been made for this purpose. The same number of sections as before are then cut out to each of the inclined corrected water-lines, the paper model prepared and the centre of buoyancy obtained as already described.

Through this new centre of buoyancy a line is drawn perpendicular to the inclined water line, and the distance between this line and the centre of gravity of the ship, already obtained, is the righting arm. If this process is repeated for each angle of inclination, it is thus seen a complete curve of stability may be approximately obtained.

A further method of arriving at results by experiment, involving principles not unlike those of the " paper section " method just described, has recently come under the author's

notice, and through the courtesy of its inventor—Mr. John H.
Heck, of Lloyd's surveying staff at Newcastle—the following

FIG. 16.

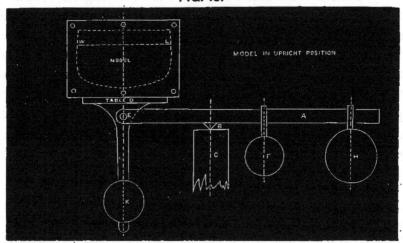

general description of the apparatus and fundamental principles
is made public for the first time :—

By means of a "stability balance," roughly i'lustrated by Figs 16 and 17, in
conjunction with either an outside or inside model of the Vesse¹, the moments of
stability can be practically determined.   In practice, an inside model has been

FIG. 17.

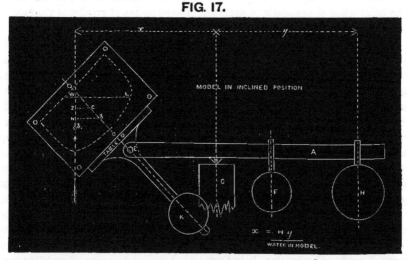

found the most convenient to employ.   This consists of a number of rectangular,
pieces of yellow pine of any uniform thickness, out of which a portion has been
cut, respectively to the form of the vessel at equidistant intervals of say 15 feet.
These pieces, together with two end pieces, are kept together by four or six bolts,
thus forming a contracted model, the inside of which is of a similar form to

that of the vessel. If this model is filled with water to a height corresponding to any draught, it will represent a volume of water having the same form, and proportional to the displacement of the vessel at that draught.

The stability balance consists of a frame A attached to a steel bar Z, having knife edges working upon the support C; a table D attached to a spindle working freely in the bearings E, and capable of being turned through any angle; a sliding weight F to balance the weight of the model when empty; a sliding weight H to balance and measure the weight of the water contained in inside or displaced by outside models; a sliding balance weight K which by adjustment will locate the centre of gravity of the combined weights of the table D, the model and the weight K in the axis of the table D, so that the model will remain when empty in any inclined position, and be balanced by the weight F.

In order to determine the moments of stability, the model is first fixed on the table D, and the weights F and K so adjusted that F will balance the model at all inclinations. The table is then brought into the upright position, and water is poured into the model to the height corresponding to the desired draught of water, and the weight H shifted until the whole is balanced The weight of water in the model will evidently be = weight H × its distance from the fulcrum ÷ distance centre of model is from fulcrum.

If the table with the model is now turned through any angle, the distance the centre of gravity of the water has moved from the axis E of the table can easily be determined by shifting the weight H until the whole is balanced, then evidently from the principles of the lever, H × by its distance from fulcrum = weight of water in model × by the distance the centre of gravity of the water in the model is from fulcrum. Since the weight of H × its distance from fulcrum ÷ the weight of water in model is known, the distance that the centre of gravity of the water has shifted from centre line is easily ascertained and the righting lever determined.

From a lengthened series of experiments, conducted by Mr. Heck—latterly in Messrs. Denny's Works where an apparatus from a special design by Mr. Heck has been constructed for the firm's use—the method gives promise of taking a firm place as an extremely simple and approximately accurate means of arriving at the stability of vessels.[*]

While a vessel's qualities with respect to stability may be determined with great precision by the naval architect, his investigations are only directly applicable to the ship while

---

[*] An obvious means of dealing approximately with stability, to which limits of space will not permit more than simple reference, consists in so manipulating the data obtained by calculation for known ships that it may be made available, either in the form of curves or of tables, for determining the stability of proposed vessels. Methods of accomplishing this may of course vary to suit the ideas and convenience of designers. A well-arranged system was brought

empty or when in certain assumed conditions of loading which may or may not often occur in actual service. He cannot for obvious reasons estimate, far less control, the amounts and positions of centre of gravity of the various items of weight that may make up the loading.* This aspect of the subject has received attention at the hands of naval architects for a considerable time, but the forcible way in which it has been brought under view by recent experience has resulted in special efforts being made to practically meet the necessities of the case. In 1877 Mr. William John read a paper before the Institution of Naval Architects, in which he dealt with the effect of stowage on the stability of vessels, and since that time such authorities as Martell, White, and Denny have given valuable papers or made suggestive comments bearing on this important matter. Much has also been done by several builders in the way of devising diagrams useful for regulating stowage and manipulating ballast with regard to initial stability. At the last meeting of the Institution, Professor Elgar read a paper on "The Use of Stability Calculations in Regulating the Loading of Steamers," distinguished by its eminently practical character, and forming an important contribution to the solution of this problem. The author disapproved of curves of stability being supplied with vessels, as had been advised and was then becoming the practice. General notes, giving in a simple form easily applied in daily practice, particulars respecting the character of a ship's stability in different conditions, are what the author recommended and had found through actual experience to meet the case most effectually. In the discussion which followed it

---

forward, jointly by Mr. F. P. Purvis, head of Messrs. W. Denny & Brothers' scientific staff, and Mr. B. Kindermann, one of his assistants, in a paper (see list at end of chapter) read before the Institution of Engineers and Shipbuilders in April last. While the results exhibited in the paper are immediately applicable to ships of one particular form, whatever the length, breadth, depth, or draught may be, this method still requires much development to make it at all universally applicable.

* It is the usual practice to assume vessels to be laden with homogeneous cargo of such a density as to fill the holds, and for this condition to estimate the position of centre of gravity to be used in calculation.

PORTRAIT
AND
BIOGRAPHICAL NOTE.

PROF. FRANCIS ELGAR.

PROF. FRANCIS ELGAR,

FELLOW OF THE ROYAL SCHOOL OF NAVAL ARCHITECTURE AND
MARINE ENGINEERING ; MEMBER OF THE COUNCIL OF THE INSTITUTION
OF NAVAL ARCHITECTS ; MEMBER OF THE INSTITUTION OF CIVIL
ENGINEERS ; AND PROFESSOR OF NAVAL ARCHITECTURE
IN THE UNIVERSITY OF GLASGOW.

BORN at Portsmouth in 1845. Received a preliminary training in practical shipbuilding, and in the drawing office, at the Royal Docykard, Portsmouth, and studied in the Mathematical School there. Was appointed an admiralty student in the Royal School of Naval Architecture and Marine Engineering, South Kensington, in 1864. In 1867 was a draughtsman and assistant surveyor, in the Admiralty Service, and in 1870 was foreman of the Royal Dockyard, Portsmouth. Left the Admiralty Service at the end of 1871 to become the principal assistant of Sir E. J. Reed, K.C.B., M.P., in the designing and surveying of war-ships, building for various Governments. In 1874 was general manager of Earle's Shipbuilding & Engineering Company at Hull. From 1876 to 1879 practised as a naval architect in London; and in 1879 went to Japan, by request of the Imperial Japanese Government, to advise upon matters relating to their navy. In 1880 visited the principal arsenals and workshops of China, and returned to this country in 1881. Since then has practised in London as a Consulting Naval Architect and Engineer, and designed and superintended the construction of numerous vessels. At the request of the builders and owners respectively, investigated the causes of the disasters which befell the " Daphne " and " Austral," and gave evidence respecting the same at the official inquiries, held in 1883. Immediately upon the "John Elder" Chair of Naval Architecture being founded in Glasgow University, through the munificence of Mrs Elder, the University Court unanimously elected Mr ELGAR as the first Professor. In 1884 was nominated by the Council of the Institution of Naval Architects as their representative upon the Board of Trade Load-line Committee. Is the author of an illustrated work upon " The Ships of the Royal Navy," and of papers read before the Royal Society and Institution of Naval Architects ; and was formerly sub-editor of the Quarterly Magazine " Naval Science."

was intimated by Mr. William Denny that his firm had already
resolved to furnish every new steamer produced by them with a
volume containing general and special notes and diagrams dealing not only with stability but with several other important
technical properties (see foot-note, page 59). After consultation
with Professor Elgar, however, he had abandoned his intention
of supplying stability curves.

An arrangement designed to readily find the position of the
centre of gravity experimentally by inclining, and to indicate
at once the stability of loaded vessels as represented by metacentric height, has been devised and introduced on board several
ships by Mr. Alexander Taylor, of Newcastle—already referred
to in connection with the triple-expansion principle in marine
engines. The instrument and apparatus, which he appro-
priately names the "Stability Indicator," was described in a
paper read by him before the Institution of Naval Architects at
its last meeting. When once an inclining operation has been
made, the degree of inclination is read from a glass gauge and
the position of centre of gravity and corresponding metacentric height from a previously prepared scale set up alongside the gauge, or from tabulated figures.

The advance made within recent years in connection with
steam propulsion comprises many matters necessarily left
unconsidered in the chapter on speed and power of modern
steamships. Scientific methods have undoubtedly contributed
in no small degree to the realization of the remarkable results
therein outlined. The achievement of one triumph after another as demonstrated in the actual performances of new
vessels, and especially the confidence with which pledges of
certain results are given and received long before actual trials
are entered upon—and that sometimes with regard to ships embodying very novel features—are evidences of the truth of this.

The oldest method of approximating to the horse-power
required to propel a proposed vessel at a given speed is to
compare the new ship with ships already built by the use of
formulae known as "co-efficients of performance" deduced

from the results of their speed trials. Two such co-efficients have been deduced from Admiralty practice, the one involving displacement, the other area of mid-section, with speed as the variable in both cases. Another method which has been largely used, consists in first determining the ratio of the indicated horse-power to the amount of "wetted surface," or immersed portion of the vessel's skin, in the exemplar ship, and then estimating from this ratio the probable value of the corresponding ratio for the proposed ship at her assigned speed. Inasmuch as these methods of procedure do not take account of the *forms* of the hulls, and consequently of that factor in the total resistance due to *wave-making*, they cannot be used with any degree of confidence, or without large corrections, except in connection with vessels whose speeds are moderate in proportion to their dimensions: those in fact in which the resistance varies nearly as the square of the speed. A further method, somewhat resembling the one based upon the relation between indicated horse-power and the "wetted surface," was proposed by the late Prof. Rankine, but has never been extensively employed. Apart from the unreliable nature of the results which an application of it gives—except for certain speeds—it is open to several serious objections in practice.

A method of analysis and prediction, meeting with considerable acceptance from shipbuilders on the Clyde and elsewhere, has been introduced within recent years by Mr. A. C. Kirk, of Messrs R. Napier & Sons.* The method consists in reducing all vessels to a definite and simple form, such as readily admits of comparison being made between their immersed surface, length of entrance and angle of entrance and their indicated horse power, and from this judging of the form and proportions best suited to a given speed or power in proposed vessels. The form in question consists of a block model, having a rectangular midship section, parallel middle body, and wedge-shaped ends; its length being proportioned to that of the ship, its

---

* See paper by Mr. Kirk "On a Method of Analysing the Forms of Ships and Determining the Lengths and Angles of Entrance."—Trans. Inst. N.A., vol. xxi., 1880.

depth to the mean draught of water, its girth of mid-section to the girth of immersed mid-section of the ship, and the surface of its sides, bottom and ends, to the immersed surface of the ship. By finding from one or more exemplar ships—the selection of which is obviously governed by the conditions of analysis—the rate of indicated horse-power required per unit of wetted surface at the speed assigned for the proposed vessel; the appropriate rate for the latter may easily be determined.

The data afforded by the modern system of progressive speed trials, especially when taken in conjunction with that of experiment with models as systematised by Mr. Froude, supplies in a reliable way much of what is most lacking in the older methods of comparison and prediction. Progressive speed trials on the measured mile were first systematically instituted by Mr William Denny about nine years ago, since which it has been the practice of his firm to make such trials with all their vessels. The practice has been followed by other firms on the Clyde and elsewhere, and there is every probability it will be still more widely adopted in the future. The system consists in trying the vessel at various speeds, ranging from the highest to about the lowest of which she is capable. The several speeds are the mean of two runs—one run with the tide and one against, the object being to eliminate the tide's influence from the results.*

Essentially noteworthy in connection with the system is the manner in which the data obtained from the trials is recorded for future use. This consists of a series of curves, representing

---

* With the view of effecting an economy in time, and to enable the trials at progressive speeds to be carried out while vessels are in a lengthened run out to sea, a method has been proposed by Mr. J. H. Biles, naval architect to Messrs, J. & G. Thomson, and adopted on board the vessels tried by that firm, and also experimented with on some of the vessels turned out by Messrs. W. Denny & Bros., by which the necessity for running with and against the tide on the measured mile is entirely obviated. The principle of the method is to measure the time that a certain part of the length of the ship takes to pass an object thrown from the bows of the vessel well clear of the side. For full particulars, both of the apparatus employed and of the results of actual trials by this method compared with trials made on the measured mile, see paper on "Progressive Speed Trials," by Mr. Biles, in the Transactions : Institution of Naval Architects, vol. xxiii., 1882.

the chief properties of ship, engines, and propeller—*e.g.*, "speed and power," "revolutions" and "slip"—which show to the eye, more easily and clearly than bare figures, the whole course and value of a steamer's performances. For that of speed and power the various speeds made at the trials are set off to convenient scale as horizontal distances, and the indicated horse-power corresponding to those speeds are set off to scale as vertical distances. The intersection of the offsets so made, give spots for the curve. The other curves alluded to are similiarly constructed, the requisite data being the direct or deduced results of the measured mile trials.

From the accumulation of trial results thus graphically recorded the designer of new ships can proceed to estimate with greater assurance of attaining satisfactory results than by employing the older methods. If, for example, a ship is to be built of virtually similar dimensions and form to one for which such information is available, but of less speed, the task is simply one of measurement from the curves, with some allowance for probable differences in the constant friction of the engines. If the speed is to be greater than that of the exemplar ship, but still within the limits when wave-making resistance assumes relative importance, the case'is also one of simple reading from the curves, with slight corrections. When both the speed and size are different, but the form is approximately the same, the case is more difficult, but it can be dealt with approximately by employing the "law of comparison" or of "corresponding speeds" enunciated by Mr. Froude. Formulae based upon this law—which will be more fully referred to presently—have been devised by one or two designers, and applied by them to problems of the latter class as they occurred in the course of their professional work. Mr. John Inglis, junr., described a method of analysis he had adopted, involving the use of Mr. Froude's law, in a paper read before the Institution of Naval Architects in 1877.

When unusual speeds are aimed at, or when novel types of vessels have to be dealt with, the only available method of making a trustworthy estimate of the power required lies in

PORTRAIT
AND
BIOGRAPHICAL NOTE.
———
WILLIAM DENNY.

WILLIAM DENNY, F.R.S.E.,

MEMBER OF COUNCIL OF THE INSTITUTION OF NAVAL
ARCHITECTS, MEMBER OF THE INSTITUTION OF CIVIL ENGINEERS,
OF THE INSTITUTION OF MECHANICAL ENGINEERS, OF THE
IRON AND STEEL INSTITUTE, AND OF THE INSTITUTION
OF ENGINEERS AND SHIPBUILDERS IN SCOTLAND.

ELDEST son of Mr Peter Denny, head of the old-established firm of William Denny & Bros., Leven Shipyard, Dumbarton. Mr DENNY was born at Dumbarton in 1847, and was educated at the High School of Edinburgh, under the late Mr John Carmichael, one of its most distinguished teachers. In his seventeenth year, he left the High School, and entered on a course of practical training as a shipbuilder in Leven Shipyard, serving for stated terms in the various departments. Since 1870 he has been a partner, and of late the managing partner, in the shipbuilding firm, and he has also shared in the partnership of the separate engineering business of Messrs Denny & Company. In addition to discharging the many arduous duties pertaining to his business position, Mr DENNY is enabled to take a prominent part in the proceedings of several of the professional societies with which he is connected. His whole theoretical training has been acquired in business, his previous education having been of a purely classical nature. In Mr DENNY this experience has been eminently fruitful of results, evidence of which may be seen in the part he has taken—both personally and as representing his firm—in various important movements dealt with in the present work. Early in the present year, on a Committee being formed by the Board of Trade to enquire into the subject of the Load Line of Vessels, Mr DENNY was appointed a member.

the use of direct or deduced results from model experiments. Mr. Froude began the work of speed experiments with ships' models on behalf of the Admiralty at the Experimental Tank in Torquay about 1872, carrying it on uninterruptedly until his death in May, 1879. Since that lamented event the work has been continued with most gratifying results by his son, Mr. R. E. Froude. Experiments had, of course, been made by many other investigators previous to Mr. Froude, but none before or since have made model experiments so practically useful and reliable. Since the value of the work carried on at Torquay has become appreciated, several experimental establishments of a similar character have been instituted. The Dutch Government, in 1874, formed one at Amsterdam, which, up till his death in 1883, was under the superintendence of Dr. Tideman, whose labours in this direction were second only to those of the late Mr. Froude. It is now superintended by Mr. A. J. H. Beeloo, Chief Constructor, and under him by Mr. H. Cop. It was here, it may be remembered, that experiments were made with a model of the Czar of Russia's yacht *Livadia*, previous to the construction of that extraordinary vessel being begun by Messrs. Elder & Co. On the strength of the data so obtained, together with the results of the trials made on Loch Lomond with a miniature of the actual vessel, those responsible for her stipulated speed were satisfied that it could be attained. The actual results as to the speed of the novel vessel amply justified the reliance put upon such experiments. In 1877 the French naval authorities established an experimental tank in the dockyard at Brest, and the Italian Government have formed one in the naval dockyard at Castellamare. The only experimental tank hitherto established by a private mercantile firm is that in the shipyard of Messrs. Denny, Dumbarton. This establishment is on a scale of completeness not surpassed elsewhere, and is fitted with every appliance which the latest experience in such experiments shows to be advantageous. A special staff of experimentalists, forming a branch of the general scientific body, are engaged conducting experiments and accumulating data, which, besides being of service in their present daily practice, must ultimately yield fruit of a very special kind to this enterprising firm.*

From mathematical reasoning, and by means of an extended series of experiments with models and actual ships, Mr. Froude determined that for two vessels of similar form—for instance a ship and her model—the "corresponding speeds" of ship and model are to one another as the square roots of the similar dimensions, and at corresponding speeds the resistance of ship and of model are to one another as the cubes of the similar dimensions—subject to a correction concerned with skin friction necessitated by the difference in the lengths of ship and model.†
Having obtained the resistance of a model, and from it, by an application of the above law, deduced the resistance of the full-sized vessel, the effective horse-power is found by multiplying the resistance by the speed of the vessel in feet per minute, and dividing by 33,000. From the effective horse-power an estimate of the indicated horse-power required can be made by using ratios which the one bore to the other in former ships, as obtained from a comparison of their model experiments with their measured mile trial results.

The value of progressive speed trials and of experiments with models as affording convenient means whereby analysis

---

* A general outline of the operations conducted in Messrs. Denny's tank will be found in the description of their large works in Chap. VI. For a detailed account of the *modus operandi* in the same establishment, see abstract of a paper delivered in Dumbarton by Mr. E. R. Mumford, of Messrs. Denny's Experimental Staff, printed in the *Engineer* for 15th February and the *Steamship* for 15th February of the present year.

† From experimental data obtained by Mr. Froude, this correction can be made with certainty. The reasons for it may be explained as follows :—If an extremely thin short plane is drawn through the water it meets a certain resistance due entirely to surface friction; that is, supposing the plane to be thin enough to eliminate wave-making and eddy-making. If the length of the plane is doubled while the depth is kept the same, the resistance at the same speed is not, as might at first appear to be the case, doubled accordingly. Owing to the friction of (say) the first half of the plane, the water is made to partake of the motion of the plane, so that the second half of the length, rubbing not against stationary water, but against water partially moving in its own direction, does not experience so much resistance from it. Adding a third equal length, it would have less surface friction than the second, and so on to infinity.

may be made of the several sources of expenditure of power in propelling vessels can scarcely be over-estimated.

From a study of the graphic records of progressive trials, and from model experiment results, Mr. Froude discovered a method whereby the power expended in overcoming the frictional resistance of the engines could be determined, and estimates made of the amount of power absorbed by other elements.. The method in question was communicated in full in a paper read before the Institution of Naval Architects in 1876, and has since been extensively used. Methods of analysis resulting from a simultaneous study of this subject, were also proposed by Mr. Robert Mansel, a prominent Clyde shipbuilder and noted investigator, but they failed in meeting with the acceptance which was at once accorded to Mr. Froude's propositions.*

Although the results obtained by an application of Mr. Froude's analysis to the trials of a large number of merchant vessels have undoubtedly thrown considerable light on the relative efficiency of hull and engines, and of various types of engines, still, for several reasons adduced by extended experience—most of which, indeed, were foreseen and perfectly appreciated by Mr. Froude himself—the need has been felt for some means of directly measuring the power actually delivered to the propellers by the engines when working at different speeds. One of Mr. Froude's latest inventions, the perfecting of which was not accomplished until after his death, consisted of a dynamometric apparatus designed to accomplish this important end.† The construction of the instrument was undertaken for the Admiralty, and trials were made with it on H.M.S. *Conquest* in the early part of 1880. The results of these experiments have not yet in any form been recorded, but there can be no question as to the benefit that would accrue to the profession if the Admiralty could be induced to publish these, as well as the results of other experiments with this instrument.

Experiments with actual vessels to determine dircetly the

---

* See papers by Mr Mansel, enumerated in list at end of chapter.

† For description of apparatus, see Trans. Inst. Mechanical Engineers, 1877.

relative efficiency of hull, engines, and propellors have on several occasions been undertaken. A series of trials of this nature were made in 1874 by Chief-Engineer Isherwood, U.S. Navy on a steam launch, the results of which may be found detailed in the Report of the Secretary of U.S. Navy for 1875. Similar trials have been made recently on the United States steamer *Albatros*, an interesting account of which appeared in *Engineering* of October 17 of the present year. These experiments are referred to as notable examples of what might be carried out with great advantage on other and larger vessels, although they are such, perhaps, as few single firms can well be expected to follow extensively.

The economies which may be obtained by changes in the propellers fitted to ships, and the great value of progressive speed trials as a means of measuring the effects of such changes, received most remarkable illustration in the results of the trials of H.M.S. *Iris*, carried out for the Admiralty in 1880. These showed that by simply varying the propellers—all other conditions remaining practically unchanged—the speed of the ship was increased from $16\frac{1}{2}$ to $18\frac{1}{2}$ knots per hour. Scarcely less striking improvements in the performances of vessels due to changed propellers might be found from the records of trials made with merchant vessels within recent years.

Inasmuch as measured mile trials are usually carried out when vessels are in the light or partially loaded condition, the results are far from being so valuable as they might be made; alike for the purposes of the naval architect, the shipowner, and ships' officers; if they were undertaken with vessels in the completely laden condition. The information obtained from the trials of incompletely laden vessels does not yield that knowledge of a vessel's qualities under the conditions necessarily imposed by actual service, which, if possessed by naval architects, would doubtless prove of immense value, nor does it furnish that standard of comparison for performances at sea which owners and captains should possess. In the interests of all concerned, it is to be hoped the practice of trying loaded vessels may become more common.

Amongst the earliest and most notable investigations involving the application of principle to the calculation of the longitudinal strength of iron vessels were those by Sir William Fairbairn, who contributed an elaborate statement of his views and methods to the first meeting of the Institute of Naval Architects in 1860. Investigation up till about this period, almost wholly concerned itself with vessels considered as girders, and in assumed conditions of fixed support, such as being pivoted on rocks. Later investigations have shown these conditions to be altogether too extreme and severe when compared with the known and estimated strains which vessels are called upon to bear in ordinary service. In 1861 Mr. J. G. Lawrie, of Glasgow, in an able paper on Lloyd's rules, read before the Scottish Shipbuilders' Association,* reasoning from wave phenomenon and the probable effects attending motion in a seaway, endeavoured to deduce limits or absolute values for the extreme strains experienced by a vessel in the circumstances, the results obtained by Mr. Lawrie bearing very closely on those deduced by later investigations. The late Professor Rankine made investigations involving consideration of strains in a seaway, and formulated several valuable rules which to some extent are still accepted, although giving results which are not likely to be exceeded in any case of ordinary service.†

For the most recent advances made in this important branch of the science of naval architecture, the profession lies under indebtedness chiefly to one or two naval architects of eminent

---

* A body which shortly afterwards joined with a kindred society in forming the " Institution of Engineers and Shipbuilders in Scotland," hereafter noticed.

† Following the methods laid down in the Treatise on Shipbuilding, edited by Prof. Rankine, Mr. John Inglis, Pointhouse, instituted calculations in 1873 of the longitudinal strains of two steamers built by his firm, the form of the waves being assumed trochoidal. The result of these calculations—which, under Mr. Inglis' directions, were got out by Mr. G. L. Watson, subsequently distinguished as a yacht designer, and then in the employ of Messrs. Inglis—appeared in the form of curves of hogging moments in *Engineering* for 1st May, 1874. Mr. Inglis found that entering upon the work of calculation had a very decided effect in giving him clearer ideas of how distribution of weight and buoyancy affected the structure of a vessel.

ability, whose professional province for a time has lain more especially in the way of a full consideration of the subject. Sir E. J. Reed, while Chief Constructor of the Navy, and under him several Government-trained naval architects subsequently acquiring high positions, achieved much in accurate investigation of ironclad vessels of war. In 1870 the authority named read an elaborate paper before the Royal Society dealing at length with such work.* In 1874 Mr. William John, formerly under Sir E. J. Reed, but at that time Assistant Chief Surveyor to Lloyd's Register, read a valuable paper before the Institution of Naval Architects, in which he gave the results of investigations of specific cases, and of long and careful study of the general problem as concerned with merchant vessels. In this paper, Mr. John advanced the proposition that the maximum bending moment likely to be experienced on a wave crest may be taken approximately as one thirty-fifth of the product of the weight of the ship into her length. Proceeding on this assumption Mr. John's paper further gave valuable results of calculations made into the strength of a series of vessels representing large numbers of mercantile steamers then afloat.† Of this paper and the conclusions it pointed to, Mr. John, in a later paper on " Transverse and other Strains of Ships," said:—

" The investigations showed unmistakably that as ships increased in size a marked diminution occurred in their longitudinal strength, and the results caused some surprise at the time, although they might perhaps have been easily inferred from the writings of others published at an earlier period. Those results, in spite of their approximate character, impressed two conclusions strongly on my mind: firstly, that there was cause for anxiety as to the longi-

---

* The substance of this paper is contained in a series of three articles on the Strength and Strains of Ships given in " Naval Science " (vol. i. and ii., 1872-3), a high-class journal ably edited by Sir E. J. Reed, but unfortunately abandoned after the fourth year of publication.

† It should be stated that under certain circumstances of lading and support the value assigned by Mr. John for the maximum bending moment may be exceeded in merchant vessels, and that in some special classes of ships—particularly light-draught vessels in certain circumstances of lading and support— the sagging moment may prove of most consequence. Instances are indeed on record of light-draught vessels giving way completely under the excessive sagging strain brought upon them at sea.

PORTRAIT
AND
BIOGRAPHICAL NOTE.

———

WILLIAM JOHN.

WILLIAM JOHN,

FELLOW OF THE ROYAL SCHOOL OF NAVAL ARCHITECTURE AND MARINE
ENGINEERING ; MEMBER OF COUNCIL OF THE INSTITUTION OF NAVAL
ARCHITECTS; MEMBER OF THE IRON AND STEEL INSTITUTE.

BORN at Narberth, Pembrokeshire, in July, 1845. Was educated in
the Mathematical School at the Royal Dockyard, Pembroke, and
received a practical training in shipbuilding in that dockyard. Was
appointed an Admiralty student in the Royal School of Naval
Architecture and Marine Engineering, South Kensington, in 1864,
and passed out in 1867 with the diploma of Fellow of the First
Class. In 1867 was appointed a draughtsman in the department of
the Controller of the Navy at the Admiralty, and served in that
capacity till 1872, when he left the Admiralty service for that of
Lloyd's Register of British and Foreign Shipping, in which Society
he was shortly afterwards appointed Assistant Chief Surveyor. In
1881 he left Lloyd's Register to become general. manager to the
Barrow Shipbuilding and Engineering Co. (Limited), at Barrow-
in-Furness, which position he now occupies. While at the Admiralty,
distinguished himself in original scientific work in naval architecture
—notably in 1868, by constructing the first curve of stability which
was ever produced ; in 1870, by investigating the stability of H.M.S.
"Captain," and pointing out, only a few days before she was lost,
the dangers to which she was liable ; also by his calculations relating to the strength of war-ships, and constructing for them the first
curves of hogging and sagging and sheering strains. Since leaving
the Admiralty, has enhanced his high reputation for scientific skill
through his investigations into the stability and strength of mercantile
ships, and the numerous valuable papers upon these and other
subjects, which he has read before the Institution of Naval Architects, and other scientific bodies. Has devoted himself largely
and very successfully to the consideration of the principal causes
of loss of ships at sea — both of sailing vessels and steamers ;
and has given most instructive evidence in some of the principal
cases which have been enquired into in recent years. Several years
ago, when sailing ships were being frequently dismasted, made
a very lengthy and complete investigation of the circumstances
in which these casualties happened, and of their causes ; and the
same is embodied in an elaborate report upon the subject to the
Committee of Lloyd's Register. Was selected by the Committee
appointed to enquire into the loss of H.M.S. Atalanta to investigate
the stability of that vessel as an independent check upon the
official Admiralty calculations, and his report and evidence showed
conclusively that she was capsizable, and probably did capsize at sea.

tudinal strength of some very large iron steamers then afloat, and that the longitudinal strength of large ships needed on all hands the most careful vigilance and attention; and secondly, that in small vessels, and even vessels of moderate dimensions, the longitudinal strength need cause but little anxiety, because it is amply provided for by the scantlings found necessary to fulfil the other requirements of a sea-going trade."

Using the formula as to the maximum bending moment advanced by Mr. John many investigations have been made subsequently into the longitudinal strength of vessels, and this increased interest in the subject has not been without its effect on subsequent structural practice.

Mr. John followed up his investigations on the longitudinal strength of merchant vessels viewed as girders by an inquiry into the transverse and other strains of ships, and in 1877 gave a valuable paper on the subject, from which a quotation has already been made, before the Institution of Naval Architects. The results of Mr. John's inquiry were such as demonstrated the need for systematic and thorough investigation of the subtle and intricate questions involved. This subject has been matter of study at Lloyd's Register for several years, and in March, 1882, the results of inquiries conducted by Mr. T. C. Read and Mr P. Jenkins, members of the staff in London, and former students of the Royal Naval College, Greenwich, were communicated in an able paper by these gentlemen, read to the Institution of Naval Architects.

It will of course be understood that many investigations of strength are instituted not necessarily out of fear that maximum strains may not be adequately allowed for, but because the dual quality of strength-with-lightness may possibly be better attained by modifications in the arrangements of material or sufficiently met by reduced scantling. The functions and influence of the Registration Societies, already commented upon (see footnote, page 103), are such as to obviate the need for strength investigations generally, or at least are such as to discourage shipbuilders from independently instituting them. Nevertheless, some well-known shipbuilders, who are also notable investigators, amongst whom may be named Inglis, Mansel, Denny, and Wigham Richardson, have done much valuable work in this connection. Mr Denny, in particular, has vigorously devoted himself to strength analysis on the basis of Lloyd's methods of fixing scantling, and read several papers on the subject, in which strong exception is taken to present practice. The healthy criticism which such labours have enabled those making them to offer regarding the Registry systems of scantlings has not doubtless failed in influencing the legislation of the Registries.

Reverting to the subject of agencies for education in naval architecture, a few remarks are due relative to Government institutions as having hitherto failed in being of immediate service to the mercantile marine. The training given to naval architects and marine engineers at the Admiralty Schools is admirably adapted for.creating a staff of war-ship designers and expert mathematicians, such as are employed in the various departments of the Admiralty service. The course of instruction has been framed expressly with a view to this, and a very high standard of mathematical knowledge is necessary before students can enter upon it. The principle of requiring one to become a first-class mathematician before attempting to teach him much of the science of naval architecture and its application in practice, is of questionable merit: at any rate it cannot be carried out in the mercantile marine. Again; economy of time and of cost of production are conditions which largely govern the methods followed in mercantile practice. Short methods of calculation, or of tentative approximation, for the purpose of enabling tenders to be made for proposed vessels, and of quickly proceeding with the work when secured, form no inconsiderable feature in the training required by mercantile naval architects. These, however, do not as a rule enter to any extent into Admiralty modes of procedure.

The want of satisfactory means for obtaining a sound scientific and practical training in mercantile naval architecture has for some time been felt to be very pressing. The evening classes conducted in most of the shipbuilding centres under the auspices of the Science and Art Department, South Kensington, are fitted to supply a part of this want so far as elementary teaching is concerned. Until recently the antiquated character of the questions set for examination was subject of general complaint, both on the part of students and teachers. In August, 1881, Mr. William Denny read a paper on " Local Education in Naval Architecture" before the Institution of Naval Architects, in which adequate expression was given to these complaints, and at the same time proposed amendments offered. As a consequence of this paper, and of the steps taken by the Institution in appointing a deputation to wait upon the Government, the questions have been considerably improved, and are now so framed as to form a fairly crucial test of a young student's knowledge of the science and practice of modern shipbuilding.

During the past three years efforts have been made by the Council of the Institution of Engineers and Shipbuilders in Scotland* to supply more adequate means of advanced education. In 1880,the Council had before them a project, promoted, for most part independently, by Mr. Robert Duncan and others, to establish a Lectureship of Naval Architecture and Marine Engineering. It was proposed to collect funds sufficient to

---

* The Institution of Engineers and Shipbuilders in Scotland was formed in 1865 through the amalgamation of two separate bodies—"The Institution of Engineers in Scotland" and "The Scottish Shipbuilders' Association." The former of these was founded in 1857 and the latter in 1860, the same year in which "The Institution of Naval Architects" was established. The membership of the Institution at the present time numbers nearly seven hundred, and comprises honorary members, members, associates, and graduates: the latter being a special section of the Institution, designed to embrace students or apprentices in the profession, and fulfilling a very useful end. The various offices have long been filled by gentlemen more or less actively engaged in the practice of shipbuilding or of engineering on the Clyde, and the proceedings have assumed, on this account alone, a richer practical interest. Scientific subjects have also received their share of attention, and of the members taking the lead in this connection the names of Mr. J. G. Lawrie and Mr. Robert Mansel are worthy of special mention. Along with Mr. Robert Duncan and Mr. Lawrance Hill, these gentlemen have, from the foundation of the Institution, taken a specially warm interest in its prosperity, and have contributed not a little thereto by the numerous valuable papers they have brought before its meetings. The secretary of the Institution is Mr. W. J. Millar, C.E., himself the author of numerous papers, and the editor of the Transactions.

endow the lectureship under the auspices of the University, and promises of substantial aid were obtained from several members. Mr. J. G. Lawrie volunteered to give the first course of lectures and did so, according to arrangement, during the winter months of 1881-82 before a considerable number of students, the lectures being delivered in the University of Glasgow during the day, and repeated in the Institution rooms in the evening. These praiseworthy efforts were still being carried on when, in November, 1883, the gratifying announcement was made of a gift of £12,500 by Mrs. John Elder, widow of the late eminent engineer, for the endowment of a Chair of Naval Architecture in the University. The founding of this chair, and the subsequent election by the University Court of Mr. Francis Elgar to the Professorship, have thus doubtless obviated the need for further efforts to found the lectureship, but there are many commendable objects connected with the University Chair to which the continued efforts of the gentlemen who supported the lecture project might fittingly be directed. Many students who can afford it will doubtless study the higher branches of naval architecture at Glasgow University, and if a few small University scholarships were established, for which all classes of workers in the shipyards and drawing offices might compete, the highest professional training would then be within the reach of the poorest of lads.

Evidences have recently been given of a strong desire on the part of many engaged in the shipbuilding and engineering industries of the Tyne and Wear for the founding of a Chair of Naval Architecture in some educational institution in that district. Along with this movement a desire has been shown for the establishment of an Institution of Engineers and Shipbuilders such as has been so long carried on successfully in the Clyde district. Definite steps are about to be taken for the realisation of these important objects, and doubtless no great time will elapse before they are accomplished.

List of papers and lectures dealing with scientific problems in shipbuilding, to which readers desiring fuller acquaintance with the *technique* and details of the subjects are referred:—

THE PROGRESS OF SHIPBUILDING IN ENGLAND: *Westminster Review*, January, 1881.

HISTORY OF NAVAL ARCHITECTURE. Lecture delivered by Mr Wm. John at Barrow-in-Furness : *Iron*, Dec. 8th, 1882.

DISPLACEMENT AND CARRYING CAPABILITY.

ON A METHOD OF OBTAINING THE DESIRED DISPLACEMENT IN DESIGNING SHIPS, by Mr R. Zimmerman : Trans. Inst. N.A., vol. xxiv, 1883.

ON FREEBOARD, by Mr Benjamin Martell : Trans. Inst. N.A., vol. xv., 1874.

ON THE LOAD DRAUGHT OF STEAMERS, by Mr W. W. Rundell : Trans. Inst. N.A., vol. xv., 1873 : vol. xv., 1874 ; and vol. xvi., 1875.

ON THE LOAD LINE OF STEAMERS, by Mr John Wigham Richardson : Trans. Inst. N.A., vol. xix., 1878.

ON THE BASIS FOR FIXING SUITABLE LOAD LINES FOR MERCHANT STEAMERS AND SAILING SHIPS, by Mr Benjamin Martell: Trans. Inst. N.A., vol. xxiii., 1882.

ON THE ASSESMENT OF DECK ERECTIONS IN RELATION to FREEBOARD, by Mr H. H. West, vol. xxiv., 1883.

TONNAGE MEASUREMENT, MOULDED DEPTH, AND THE OFFICIAL REGISTER IN RELATION TO THE FREEBOARD OF IRON VESSELS, by Mr W. W. Rundell : Trans. Inst. N.A., vol. xxiv., 1883.

STABILITY.

ON THE CALCULATION OF THE STABILITY OF SHIPS AND SOME MATTERS OF INTEREST CONNECTED THEREWITH, by Mr W. H. White and Mr W. John : Trans. Inst. vol. xii., 1871.

ON THE RELATIVE INFLUENCE OF BREADTH OF BEAM AND HEIGHT OF FREEBOARD IN LENGTHENING OUT THE CURVES OF STABILITY, by Mr Nathaniel Barnaby : Trans. Inst. N.A., vol. xii., 1871.

ON THE LIMITS OF SAFETY OF SHIPS AS REGARDS CAPSIZING, by Mr C.W. Merrifield : *The Annual* of the Royal School of Naval Architecture and Marine Engineering, No. 1, 1871 ; London, H. Sotheran & Co.

ON CURVES OF BUOYANCY AND METACENTRES FOR VERTICAL DISPLACEMENTS, by Mr George Stanbury : *The Annual* of the Royal School of Naval Architecture and Marine Engineering, No. 2, 1872 , London, H. Sotheran & Co.

THE GEOMETRICAL THEORY OF STABILITY FOR SHIPS AND OTHER FLOATING BODIES : *Naval Science*, vol. iii., 1874, and vol. iv., 1875 (Three Articles).

ON THE METACENTRE AND METACENTRIC CURVES : *Naval Science*, vol. iii., 1874.

ON POLAR DIAGRAMS OF STABILITY, by Mr J. MacFarlane Gray: Trans. Inst. N.A., vol. xvi., 1875.

ON THE STABILITY OF SHIPS, by Mr Wm. John: Trans. Inst. N.A., vol. xviii., 1877.

ON THE GEOMETRY OF METACENTRIC DIAGRAMS, by Mr W. H. White : Trans. Inst. N.A., vol. xix., 1878.

ROLLING.

RESISTANCE, SPEED, AND POWER.

ON STREAM LINE SURFACES, by Prof. W. J. Macquorn Rankine : Trans. Inst. N.A., vol. xi., 1870.

ON EXPERIMENTS WITH H.M.S. GREYHOUND, by Mr William Froude : Trans. Inst. N.A., vol. xv., 1874.

ON THE DIFFICULTIES OF SPEED CALCULATION, by Mr Wm. Denny : Trans. Inst. Eng. and Ship. in Scotland, vol. xvii, 1874-75.

ON THE RATIO OF INDICATED TO EFFECTIVE HORSE POWER AS ELUCIDATED BY MR DENNY'S MEASURED MILE TRIALS AT VARIED SPEEDS, by Mr Wm. Froude : Trans. Inst. N.A., vol. xvii., 1876.

ON THE COMPARATIVE RESISTANCES OF LONG SHIPS OF SEVERAL TYPES, by Mr Wm. Froude : Trans. Inst. N.A., vol. xvii., 1876.

ON EXPERIMENTS UPON THE EFFECT PRODUCED ON THE WAVE-MAKING RESISTANCE OF SHIPS BY LENGTH OF PARALLEL MIDDLE BODY, by Mr Wm. Froude : Trans. Inst. N.A., vol. xviii., 1877.

ON STEAMSHIP EFFICIENCY, by Mr Robert Mansel : Trans. Inst. Eng. and Ship. in Scotland, vol. xxii., 1878-79.

ON THE TRUE NATURE OF THE WAVE OF TRANSLATION AND THE PART IT PLAYS IN REMOVING THE WATER OUT OF THE WAY OF A SHIP WITH LEAST RESISTANCE, by Mr J. Scott Russell : Trans. Inst. N.A., vol. xx., 1879.

ON THE LEADING PHENOMENA OF THE WAVE-MAKING RESISTANCE OF SHIPS, by Mr R. E. Froude : Trans. Inst. N.A., vol xxii., 1881.

MR. FROUDE'S EXPERIMENTS ON RESISTANCE AND ROLLING : *Naval Science*, vol. i., 1872, and vol. iv., 1875.

MR. FROUDE'S RESISTANCE EXPERIMENTS ON H.M.S. GREYHOUND : *Naval Science*, vol. iii., 1874.

ON A METHOD OF RECORDING AND COMPARING THE PERFORMANCES OF STEAMSHIPS, by Mr John Inglis, jun. : Trans. Inst. N.A., vol. xviii., 1877.

ON A METHOD OF ANALYSING THE FORMS OF SHIPS AND DETERMINING THE MEAN ANGLE OF ENTRANCE, by Mr Alex. C. Kirk: Trans. Inst. N.A., Vol. xxi., 1880.

ON SOME RESULTS DEDUCED FROM CURVES OF RESISTANCE AND PROGRESSIVE M M SPEED CURVES, by Mr J. H. Biles : Trans. Inst. N.A., vol. xxii, 1881.

ON PROGRESSIVE SPEED TRIALS, by Mr J. H. Biles : Trans. Inst. N.A., vol. xxiii., 1882.

STRUCTURAL STRENGTH.

THE DISTRIBUTION OF WEIGHT AND BUOYANCY IN SHIPS: *Naval Science*, vol. i., 1872.

THE STRAINS OF SHIPS IN STILL WATER : *Naval Science*, vol. i., 1872.

THE STRAINS OF SHIPS IN EXCEPTIONAL POSITIONS ON SHORE : *Naval Science*, vol. ii., 1873.

THE STRAINS OF SHIPS AT SEA : *Naval Science*, vol. ii., 1873.

ON THE STRENGTH AND STRAINS OF IRON SHIPS : *Naval Science*, vol. iii., 1874.

ON THE STRENGTH OF IRON SHIPS, by Mr William John : Trans. Inst. N.A., vol. xv., 1874.

On Useful Displacement as Limited by Weight of Structure and of Propulsive Power, by Mr Wm. Froude : Trans. Inst. N.A., vol. xv., 1874.

On the Modulus for Strength of Ships, by Mr J. MacFarlane Gray : Trans. Inst. N.A., vol. xvi., 1875.

On the Strains and Strength of Ships, by Mr John Wigham Richardson : Trans. Inst. N.A., vol. xvi., 1875.

On Transverse and other Strains of Ships, by Mr William John : Trans. Inst. N.A., vol. xviii., 1877.

On the Strains of Iron Ships, by Mr William John : Trans. Inst. N.A., vol. xviii., 1877.

On Lloyd's Numerals, by Mr William Denny : Trans. Inst. N.A , vol. 1877.

On Lightened Scantlings, by Mr Wm. Denny: Trans. Inst. N.A., vol. xix., 1878.

On the Effect of Depth upon the Strength of a Girder to Resist Bending Strains, by Mr Frank P. Purvis : Trans. Inst. N.A., vol. xix., 1878.

On an Application of the Decimal System of Measurement in Practical Shipbuilding, by Mr Henry H. West: Trans. Inst. N.A., vol. xix., 1878.

On Longitudinal Sea Strains in Vessels as Indicated by Lloyd's Experience, by Mr Robert Mansel : Trans. Inst. Eng. and Ship. in Scotland, vol. xxi., 1877-78.

On the Strength of Iron Vessels, by Mr Geo. Arnison, jun.: Trans. Inst. Eng. and Ship., vol. xxii., 1878-79.

Freeboard and Displacement in Relation to Strains in Ships Among Waves, by Mr W. W. Rundell : Trans. Inst. N.A., vol. xxii., 1881.

On the Transverse Strains of Iron Merchant Vessels, by Mr P. Jenkins and T. C. Read :' Trans. Inst. N.A., vol. xxiii, 1882.

On Hogging and Sagging Strains in a Seaway as Influenced by Wave Structure, by Mr W. E. Smith : Trans. Inst. N.A., vol xxiv., 1883.

EDUCATION IN NAVAL ARCHITECTURE.

On the Course of Study in the Royal Naval College, Greenwich, by Mr W. H. White : Trans. Inst. N.A., vol. xviii, 1877.

On the Royal Naval College and the Mercantile Marine, by Mr Wm. John : Trans. Inst. N.A., vol. xix., 1878.

On Local Education in Naval Architecture, by Mr William Denny : Trans. Inst. N.A., vol. xxii , 1881.

CHAPTER V.

SINCE the early days of iron shipbuilding, when hand labour entered largely into almost all the operations of the shipyard, the field of its application has been gradually narrowed by the employment of machinery. The past few years have been uncommonly fruitful of changes in this direction, and many things point to the likelihood of manual work being still more largely superseded by machine power in the immediate future. Such changes, however, have not, as might be assumed, had any very sensible effect in diminishing the number of operatives generally employed. The influence has rather been absorbed in the greatly increased rate of production, and the elaboration and enhanced refinement of detail demanded by the much more exacting standard of modern times. The need for skilled handicraftsmen may not now be so general, but the skill which is still indispensable is of a higher character, and has called into existence several almost entirely new classes of shipyard operatives.

The extended employment of machinery has given impetus to, and received impetus from, the system of "piece-work" now so much in vogue in shipyards. In several of the operations, such as riveting and smithing, the nature of the work peculiarly lends itself to the system, and piece-work has consequently been in force, as regards these operations, for many years. In several other departments, however, such as plate and bar fitting, joinery, and carpentery, piece-work is only contemporaneous with and largely the consequence of improved modern machinery. Reference to "piece-work" here is not made with the intention of discussing its effects on the

labour question—concerned as this is with such large issues—
but simply of showing what effect the system has had on the
character of shipyard workmanship.   It was a favourite argu-
ment some years ago, when piece-work was being rapidly
extended, that the system was bad because it would lead to
and foster scamp-work and bad workmanship.   The results of
the past dozen years' experience disprove this completely, and
for reasons which, as early as 1877, were pointed out by Mr
William Denny—to whose spirited advocacy and adoption of
the system its present degree of acceptance with workmen is in
no small measure owing.   In his admirably written pamphlet
on "The Worth of Wages," published in the year named, Mr
Denny says: —

" As to piece-work leading to bad workmanship, this would certainly be the
result were no special arrangements made to prevent it.   These special arrangements include a right system of inspecting the work, and the rejection, at the
workman's cost, of all bad and inferior work.   There is no difficulty in carrying
out such a system, for foremen, freed from the necessity of watching the quantity of the work which is looked after by a special clerk and of checking the
laziness of their men, can give their whole attention to the matter of quality.
In fact, piece-work compels so thorough an inspection, that we find the work
done under it in our iron department much superior to what used to be done
some years ago on time.   It is very curious that trades' unionists never have
been very anxious as to the quality of their work till they had piece work to
contend with, and I have never known workmen produce such good work, as
after a few experiences of having their workmanship condemned for its bad
quality, and the cost taken out of their pockets.   Under the old time wages no
such effective stimulus urged a man on to make his piece of work up to a
proper standard."

What was true of the system as exemplified in Messrs
Denny's experience previous to 1877, holds equally good for
all the yards in which piece-work is now the rule.   Under it
work is done quicker and better than by the old system, and
so popular is it amongst workmen that a deep-rooted dislike
for "time-work" prevails where piece-work has once been
instituted and efficiently managed.

The machines in use at the present day for preparing the
separate and multitudinous pieces of material which go to form
the hull structure of iron and steel vessels are both numerous
and highly efficient.   This work of preparing material, it may

be shortly stated, mainly consists of shearing and planing the edges of plates and bars—these as supplied by the manufacturers being, of course, only approximately near the final form and dimensions—rolling and flattening or giving uniform curvature to plates; bending angle or other bars, such as are used for deck beams; and punching the holes through plates and bars for the reception of rivets. In this list regard is not had to the operations concerned with material in the heated state, the features requiring to be thus manipulated being mainly the frames of the vessel; the work being effected without the aid of any special machine tools. A small proportion of the plating also requires to be operated upon in this state, and for this purpose machine tools are sometimes brought into requisition, some notice of which will be taken further on.

While most of the machines have been introduced for a period exceeding that with which our review is more directly concerned, improved types have been made, and entirely new machines brought into requisition during recent times. The universal adoption of piece-work in almost all the departments of construction has demanded a more economical type of machine than formerly. In this way punching machines, which play so important a part in shipyards, have risen from a working speed of about fourteen rivet holes per minute to thirty and even—in the case of frame punching—to as high as forty per minute. Other machines have had a corresponding increase in speed; in several of the best appointed yards the general increase being about sixty per cent.

The introduction of the double bottom for water ballast in ships, brought about a great increase in the amount of necessary punching caused by the numerous man-holes required through the floors and longitudinals. These man-holes, oval in shape as shown by Fig. 1—of say 18-ins. by 12-ins.—had to be punched all round by the rivet-punch, and the edges afterwards dressed by hand with a chisel. To economise work in this connection, need was felt for a machine which would be capable of punching a man-hole of the ordinary size out of the thickest plate at one operation. In 1879, at the

request of one of the prominent Clyde firms, Messrs Craig &
Donald, the well-known machine-tool makers of Johnstone,
introduced a man-hole punching machine which cut holes
18-ins. by 12-ins. at the rate of seven per minute, in such a
way that no after-dressing with chisels was required. This
machine, an ordinary eccentric motion one driven by its own
engine, although tested and found capable of cutting an 18-in.
by 12-in. hole through a plate 1-in. thick, was superseded in
the yard for which it was made, by another, designed to meet
the requirements of the heaviest type of vessels built on the
cellular principle. This machine—also made by Messrs Craig
& Donald, and five or six of which are now at work in yards
on the Clyde and at Barrow—was capable of piercing a hole
30-ins. by 21-ins. through a plate ¾-ins. thick, at one operation,
and was actuated by hydraulic power. The ordinary eccentric
machine, driven by engine attached, is still in favour for
lighter work, and machines of this type are at work in several
of the East Coast yards capable of punching holes up to 21-ins.
by 15-ins. through plates ¾-ins. thick.

Reverting to the subject of the proportion of material requiring to be heated before manipulation, it is noteworthy that the
employment of mild steel is a source of economy in this connection as well as in the many others already noticed. The
superior homogeniety and great ductility of the material
favours cold-bending when such an operation would be fatal
to iron. Not only does an economy in labour result, but
incidentally there is a further advantage. Cold-bending dis-
tresses steel less than hot-bending, and the special precautions
so often taken, in the way of annealing, to toughen steel which
has been operated upon when hot, are thus obviated.

A certain proportion of the bottom plates in a ship—*e.g.*,
those adjoining the keel—and a few at the stern and elsewhere,
have quick bends and twists which are much more difficult to
treat than the easy and generally uniform curvatures on the
plates of the bilge. The latter are effected in great measure
by the "bending rolls" with the plates perfectly cold, but the
former have to be made with the plate in the heated state.

Hydraulic presses have been used for this purpose for some years, a certain proportion of the work done being the manipulation of plates while cold. With steel as the material to be operated upon, these machines are being more and more utilised in this direction, and their presence in the shipyard, as in boiler works, is sure to become more and more prevalent. The operations of the shipyard, in short, have been gaining in exactitude every year, and have borrowed both in the matters of methods and of appliances from the marine boiler works, where machine tools are more conspicuously a feature. Machine tools for riveting, now playing so important a part in shipyards, first had their utility approved in boiler shops, and the introduction of improved types of drilling machines is largely the reflected successes attending them there.

From the foregoing imperfect sketch of the principal directions in which machine tools used in *preparing material* for the constructive stage have been improved or recently introduced, it will be gathered that hydraulic power in lieu of steam has taken a prominent place in shipyards. That this is so to a remarkable extent will sufficiently appear from what follows regarding the appliances used in the work of *binding the structure* of vessels. It may, however, be premised that in several establishments hydraulic pressure has now displaced steam power in almost all the machine-tools used in the iron departments. This is so in the case of the Naval Dockyards of Toulon and Brest, in France, and of the Spanish naval establishments at Ferrol, Cadiz, &c.; the machinery in the former of which was fully described in June, 1878, before the Institution of Mechanical Engineers, by M. Marc Berrier-Fontaine, of the French Navy. The plant and machinery are by Mr. Ralph H. Tweddell, C.E., of Delahay Street, London, whose numerous inventions and great experience in this special branch of engineering are well worthy of recognition. The machines comprise those for punching, shearing, angle cutting, plate bending, and riveting, and the author referred to is high in his praise of the superior efficiency and economy of the hydraulic system, as exemplified in practice. One or two of the leading advantages of the system may be here summarised. Hydraulic machines do not consume any power at all during the interval between employment, and the power can be applied at any moment without preparatory consumption, and stopped equally quick. No shafting or belting is required, and the wear and tear of continuous motion, as in steam machines, is thus obviated. The power exerted is much more gradual than that of steam, performing the work more thoroughly, and with less liability to strain or otherwise damage the material operated upon, or the tool itself.

Although hydraulic machinery was successfully introduced by Sir William Armstrong so long ago as 1836, and has since been applied by him and others in almost every direction the application of hydraulic power to machines for constructive purposes is of comparatively modern date. Its early employment as the motive power for machine-tools was in the case of machines which were "stationary" or "fixed" in position when in use. Machines for riveting purposes in boiler shops and locomotive works were the first tools of any note to which hydraulic power transmitted from a distance was applied, but even this dates back only to about 1865. In that year Mr. R. H. Tweddell, already referred to, designed hydraulic plant, consisting of pumps, an accumulator, and a riveting machine, which were first used by Messrs Thompson, Boyd & Co., Newcastle-on-Tyne, with satisfactory results. The work was done perfectly, and at about one-seventh of the cost of hand work, and the same power was utilized in actuating hydraulic presses for such purposes as setting or "joggling" angle or tee irons. Excellence and economy of work were thus secured; and in a comparatively short time above 100 machines were at work in various dockyards and large works.

Although patent designs for portable hydraulic riveters existed before 1871, it was not till that year that any form of portable rivetters was applied in practice with any degree of success. Previous to that year the frames of ships had been riveted by Mr. Tweddell's stationary hydraulic machines, but a portable riveter invented by that gentleman in 1871 was then tried, when it was thoroughly demonstrated that during a working day of 10 hours the machine was capable of closing 1,000 rivets. Not much encouragement, however, was received from shipbuilders at the time, owing chiefly to the fact that the wages for riveting labour was not then a very urgent question. On a modification of the general plan of working, these machines being proposed by their inventor in 1876, they received more cordial recognition from shipbuilders thereafter. It is only, however, within the past five years or so that portable rivetters have been so extensively introduced into shipbuilding yards. The success which has attended them during the period leaves no reasonable doubt as to their ultimate place in every well-appointed shipbuilding establishment. Already the majority of Clyde shipyards—including all the larger ones—and most of the yards in the Tyne and Wear districts, are furnished with hydraulic riveting machines and plant, overtaking work constantly, efficiently, and with greatly reduced expense, that is matter of envy in yards not similarly favoured. In most of the larger Clyde yards the Tweddell machinery and plant are employed; but in some cases machines introduced by Mr. William Arrol, Dalmarnock Ironworks, Glasgow—chiefly for riveting the frames, beams, &c.—are used. The Arrol machines work on a similar principle to those of Mr. Tweddell, whose system is practically the only one in use on the Tyne and the Wear, and at Barrow.

The prime cost of furnishing a complete hydraulic plant is of course considerable, and such as might perhaps appear an outlay not speedily enough recouped. In view, however, of the uncertain and oftentimes harrassing conditions—not to speak of the pecuniary loss — under which the riveting department of shipbuilding work is conducted in the ordinary way, shipbuilders are constrained to acknowledge the economic advantages of the hydraulic system. Neither expense nor trouble have been spared in several yards to extend the hydraulic system into every feature where hydraulic work is practicable. The only feature now for which the machines presently in use are not available is the shell plating, and perhaps the decks, where such are entirely laid with plates. Indeed, it may fairly be said that hydraulic riveters have virtually supplanted manual riveting in nine-tenths of the structural features of a vessel. The percentage of rivets closed by machinery to the total number of rivets employed in a vessel's structure has been computed to be about fifty per cent. In one of the yards fitted with the Tweddell system the following comprise the list of structural features for which the hydraulic riveters are daily employed:—Double bottom, including the thousands of detached pieces of plates and angles of which the bracket floor style of bottom is composed; side bars attaching frames to double bottom, frames and reverse frames, beams, stiffening bars, gunwale bars, keelsons, and keels.

The shell plating, as has already been said, is about the only feature for which inventors and manufacturers of hydraulic riveters have now any serious difficulty in making provision. But many minds are exercised with the problem, and doubtless at no very distant date the present obstacles will be surmounted. One aspect of the question—and one which certain classes are apt to overlook—is that which regards the *mutual* adaptation of means to the end desired. Shipbuilders have often under consideration the practicability of so modifying structural features and methods of work as that inventors of mechanical riveters will be met half-way in supplying the much-felt desideratum. Referring to this subject, Mr. Henry H. West, chief surveyor to the Underwriters Registry for Iron Vessels, in a paper on "Riveting of Iron Ships," read before the Institution of Naval Architects at its last meeting, said:—

"May I urge upon shipbuilders the importance of endeavouring to extend the application of power riveting to the shell plating of iron vessels. By this means we shall both increase the frictional resistance, and also, by more completely filling the rivet holes, vastly improve the rigidity of the riveted joints. The difficulty of completely and exactly filling the countersink of a countersunk hole with a machine-closed rivet suggested to my friend Mr. Kirk the idea of entering the rivet from the outside, both the rivet and the countersink being made to guage, and then closing up with a machine snap-point on the inside of the ship. What progress he has made in this direction I do not know, but the difficulty does not appear to be an insuperable one. If however, we are prepared to sacrifice a fair appearance to utilitarian simplicity, there seems no sufficient reason why, above water, all the rivets should not be closed up with snap heads and points, both inside and outside. In whatever way it is accomplished, I look to the use of machine riveting as one very great step in advance in the future improvement of the riveted joints of iron ships;.and if the weight of iron vessels is to be reduced in any important degree, or if the dimensions and proportions of large merchant steamers are to increase in the future as they have done in the past, I feel sure that one of the first steps must be the reconsideration of our butt fastenings.''

The increased engine power now demanded in steamships undoubtedly points to the turther adoption of mechanical riveting—if vessels are to successfully withstand the enormous strain and vibrations to which they are thereby subject. While several have already shown drawings of the shell difficulty having been met, Mr. Tweddell, whose experience in common with that of his manufacturers and co-patentees, Messrs Fielding & Platt, of Gloucester, may justly be considered greatest in this branch of engineering, has never illustrated this. It may be mentioned, however, that excellent flush riveting is constantly done by the Tweddell hydraulic riveters, and that the same plan suggested by Mr. Kirk of entering rivets with prepared counter-sunk heads from one side, and snap pointing them by machine on the other has been long in use by Messrs Fielding & Platt. In conjunction with Mr. Tweddell, this firm have also designed several efficient arrangements to ensure the machine being kept in position until the unfinished head of the rivet is formed. Judging from these facts, there seems good reason to hope that the production of riveting machines required to overtake the remaining features will not be very long delayed.

To show that where the exigencies of the times necessitate them, expedients involving inventive skill and industrial intrepidity are never quite wanting, it may be related that several years ago, during a prolonged strike of riveters, the principal of the firm of Messrs A. M'Millan & Son, Dumbarton, introduced a portable riveting machine for the shells of ships. The machine, although improvised, as it were, to meet an emergency, fulfilled all that was expected of it, and won the approval of Lloyd's Surveyors for the Clyde district, as well as of a special deputation selected by the Committee of Lloyd's in London from among the chief surveyors of the United Kingdom. Their verdict on the performances of the machine after due inspection was that it "thoroughly fills the holes and countersinks, and produces a smoother and better clench than can usually be obtained by hand labour." From this it will be seen that in the yard of Messrs M'Millan the matter of machine riveting has received early and earnest consideration. Indeed, the extent to which hydraulic riveting is presently employed by this firm so well represents the development and progress made in this direction throughout other yards that the system adopted in their establishment may be described somewhat in detail.

The hydraulic plant and numerous different classes of portable riveters are on the Tweddell system. The hydraulic power required to work the various machines is furnished by a pair of vertical steam-engines, geared to a set of two-throw pumps, which force the water at a pressure of 1,500-lb. per square inch into an accumulator. This latter feature, as is well known, serves to store up the power in a considerable amount ready to meet the sudden demands of one or more of the riveters without calling on the pumps. As is the case in all machinery on this system, the accumulator is loaded to a pressure of 1,500-lb. per square inch. The means employed for the transmission of the water-power, from the service of main pipes laid as required throughout the yard, are flexible copper pipes, admitting of being led almost in any direction, however irregular, without being impaired or rendered inefficient. When the plant was laid down about four years ago, Messrs M'Millan determined to err if anything on the side of prudence, and they laid all their mains of double the required size, so that they could, if the high pressure was found objectionable, return to the lower pressures sometimes employed; they have, however, never found it advisable to do so.

In this yard can be seen portable riveters suspended over a vessel's deck between 40 and 50 feet above ground, capable of reaching and clenching rivets in stringers at a distance of 4 feet 6 inches from edge of plate. The power brought into play in closing some of these rivets is very great—from 20 to 30 tons —and yet this is conveyed by a small tube of only half-inch outside diameter in some cases through a distance of many hundred feet. The portable riveter here indicated is suspended on a light and handy carriage, which can travel the upper deck from stem to stern, being made purposely low so as to clear poop and bridge deck beams if such should be fitted. With this machine Messrs M'Millan have closed from 400 to 450 rivets per day of nine hours in stringers 3 feet 6 inches wide. They have also effected some very heavy work in attaching the sheer strake to the gunwale bar, the rate of progress being correspondingly satisfactory. The same features in the *Alaska,* built by Messrs John Elder & Co., were similarly operated upon by another of Mr Tweddell's riveters, whose complete system has been adopted in this large establishment also. By an elongation of the suspending arm Messrs M'Millan hope to execute, besides the stringers, most of the deck work, such as ties, diagonals, hatch coamings, &c., in one traverse of the carriage. Moreover, a second carriage with riveter may be doing simultaneously the same work on the other side of the vessel. Indeed, it only requires a further development of such work to make the riveting of complete iron decks practicable, and—with the rate of wages, for hand riveted work, usually prevailing—profitable also.

The riveting of the frames and beams is the simplest of all the work overtaken by the hydraulic riveters, and it is here the system is seen to most advantage. In any yard furnished with these machines rivets are closed at a greatly accelerated rate compared with work done by hand. Tweddell machines have been known to close, in beams, 1,800 to 1,900 rivets per machine per day of $9\frac{1}{2}$ hours. In frames the average rate at which rivets are closed is about 1,400 per day. The cost for this section of riveted work has been computed to be about one-half of that by hand, and the quality of the work is everywhere acknowledged to be better. With the same number of men the work is accomplished in something like one-third of the time. The *modus operandi* in overtaking

FIG. 22.

TWEDDELL PORTABLE FRAME AND BEAM RIVETER.

this feature of the work may be briefly described. For the riveting of the frames, in almost every case, two cranes of any convenient construction are fixed at the head of the berth in which the vessel is to be built; the frames are laid across the keel as in hand work, and rest on trestles, where the portable riveter, carried on the before-mentioned cranes, rivets them up. As the riveting in each frame is completed it is drawn down the keel by steam or hand power, and set up in place. The riveting of the beams is a still more simple operation, the beam to be riveted being placed under a gantry somewhat longer than the beam itself, and upon which the portable riveter travels. The suspending gear in this and other of the Tweddell machines combines the functions of hydraulic lifts for raising or lowering the riveter, and of conveying the necessary hydraulic pressure to the riveter. The beam is supported on trestles, and the riveter, having the facilities for travel and exact adjustment just described, accomplishes the surprising work before mentioned.

The conditions under which the riveting in cellular and bracket bottoms is accomplished are less favourable to expeditious work. This system of ship's bottom is greatly more complex in its constructive features than the ordinary bottom. The separate plates and angles which go to form the bracket floor system are to be numbered—in vessels of the average size—by thousands. The frames in such vessels are formed of three parts; one part stretches across the bottom and abuts against the plates forming the sides of the cellular bottom; the other two parts form the sides of the vessel, but are not erected until the bottom portions of the frames have been laid and all the bracket and longitudinal girders are erected and fitted upon them. On the bottom, as thus described, the portable riveters are required to operate, in many instances having to reach the rivets at a distance of 4 feet 6 inches from the edge of the plates, and in confined spaces of 24 inches. When the frames and beams are completely riveted and beginning to be erected, a travelling crane (in Messrs M'Millan's two travelling cranes are employed working from separate ends of the vessel) carrying a large portable riveter, is placed on the top of the floors, with short lengths of planking laid to act as tramways. The perfect control thus obtained is somewhat

extraordinary. The crane jib has sufficient rake to command
the whole floor of the ship, and every rivet can be closed in
the confined spaces already described. Some 800 rivets per
day can be put in, many of them at a distance of 4 feet 6 ins.
from the edge of the plate. The quality of the work is all
that could be desired; in some parts, indeed, the use of the
felt-packing necessary in hand work has been found to be
unnecessary owing to the tight work obtained by hydraulic
riveting. One crane with its riveting machine can, in a
vessel of moderate size, say 3,000 to 4,000 tons, fully keep
pace with the up-ending of the frames, provided it has something of a start. As it advances the lower deck beams are
put in place behind it, and the other work follows in order.
In ships of the more ordinary construction, longitudinal keelsons are fitted, which are readily reached by special portable
riveters, suspended by means of neat devices, some of them the
ideas or suggestions of workmen in Messrs M'Millan's service.

The only machine of the series of portable riveters employed
by Messrs M'Millan which remains to be noticed is that which
overtakes the riveting of keels. This machine is perhaps one
of the most perfect of the series, performing its functions
satisfactorily, viewed from whatever standpoint. The riveting
required on the keel of large vessels is very heavy, especially
if the through-keelson and side-bar system is adopted, when
five thicknesses of plate have to be connected, the rivets
employed being $1\frac{1}{8}$-inch or $1\frac{1}{4}$-inch in diameter. The situation
is not favourable for getting at the work to be done, the head-
room available not often exceeding $2\frac{1}{2}$ or 3 feet. These
conditions render great compactness, together with portable-
ness, necessary in the machine. The keel itself was utilised
for the attachment of the Tweddell riveter as first tried,
then again a sort of light trestle was employed, the riveter
being at one end of a lever racking on this. These plans were
abandoned, however, in favour of the machine as now used in
various yards throughout the country, an illustration of
which is given by Fig. 23. A low carriage is travelled
down alongside the keel. This carriage supports a

balanced lever, carrying at one end the riveter, capable of exerting about 50 tons on the rivet head, and at the other a balance weight. This lever can in its turn revolve horizontally about a short pillar fixed on a turn-table, thus affording unlimited control over the riveter by the man in charge; enabling him, indeed, to adjust the riveter to every irregularity of position or direction of the rivets in keel. As many as 420 1¼-inch rivets per day have been put in by this machine, an amount which is fully equal to the work of two squads of

FIG. 23.

TWEDDELL HYDRAULIC KEEL RIVETER.

riveters, and in one yard 70 rivets have been closed in as many consecutive minutes.

It may be stated generally that the several hydraulic riveters require two men to work them, and the rivets are heated in portable furnaces and dealt out in any quantities required, by a boy in attendance. The quality of the work done is superior to hand work, chiefly in that when rivets are well heated the pressure is equalised, and affects the rivets throughout their entire length, filling the holes to their utmost.

This advantage tells more in the case of keel riveting, and that it is so is evidenced by the fact, as communicated by a foreman having great experience, that rivets ¼-inch longer than rivets closed by hand have even less superfluous surface material when closed by the machine.

From the facts above detailed, taken in conjunction with the opinions of such authorities as Mr West, it can fairly be claimed for Mr. Tweddell as the inventor of the earliest of the hydraulic riveters now so extensively employed in shipyards, that he has greatly improved the character of work in ship construction. Not only so, but he has relieved the shipyard artizan from a species of work which requires little or no skill in its execution—work, indeed, which may properly be relegated to, as it certainly in course of time will be included in, that vast domain in which water, steam, electricity, and the other natural powers are so wondrously made to play their part.

While the extended use of improved machinery has brought about changes in the iron-working departments of shipyards that are structurally of the greatest importance, it is nevertheless true that the largest acquisition to shipyard machinery of late has been made in the wood-working departments. It is here, beyond question, where the equipment of modern shipyards is seen to be so much an advance on the former order of things, when handicraft was indispensable and paramount; and it is also here, probably, where the greatest labour-saving advances have been made. The artistic perfection which is evinced in the palatial saloons and state-rooms of many modern steamships would not have been possible—commercially so, at least—to the shipbuilders of twenty years ago, whose appliances, regarded from present-day standpoints, seem to have been woefully crude and meagre. Still, it is not by any means to be understood that all the shipyards of to-day are alike commentaries on the former state of things, because even now there are not wanting yards in which the necessary wood-work for ships is accomplished with singularly few machines. The need for accessions in this direction, however, is being more keenly felt every day, and in many yards quite recently the entire joinery department has been thoroughly re-organised and equipped. The chapter which follows will be devoted to descriptions of some representative establishments in the several districts, and as special references may therein be made to the machinery equipment of the wood-working departments, the present remarks will only be of a general nature.

The conversion of wood from the absolutely rough state into finished and finely-surfaced material, ready for immediate use in the interior of vessels, forms at the present time not an uncommon portion of the daily work in shipyards well equipped with modern machinery. This is not only concerned with the commoner woods employed in large quantities for structural purposes, but also to a considerable extent with those various ornamental hardwoods entering into the decorative features. The change of which this is indicative is one of increased self-dependence and economy formerly not dreamed of in ship-yards, and of improvements at every stage in the machinery for wood conversion, which are simply wonderful. In circular and straight saws, planing, moulding, and shaping machines, band and fret-saw machines, mortising, tenoning, and dove-tailing machines, and in machines for scraping, sand-papering, and miscellaneous purposes, not a few modern shipyards reflect the fullest engineering progress as concerned with wood-working machinery. In planing machines especially are the labour-saving advantages made apparent. As illustrating this it may be explained that machines of this kind in daily use are able to plane a greatly increased breadth of surface, to work several sides of the wood at one operation, and at a marvellously accelerated speed as compared with hand-work. Similarly, as regards the formation of mouldings, it may be stated that a moulding which would take a competent workman some hours to produce can be completed on a good machine in less than *one minute.* Many patterns of mouldings and other decorative items now largely used are thus only possible—commercially if not otherwise—through the extended employment of machinery. The degree of "finish" now put upon the plainest features—rendered pecuniarly possible by the use of machinery —is nowhere so striking as in the scraping of panels and the sand-papering of large surfaces. In one shipyard the author has witnessed the scraping of hardwood panels as broad as 30-ins., the shaving taken off being of marvellous thinness and perfectly uniform and entire throughout the length and breadth of panel. The surface left on the panel is beautifully smooth, rendering any after-dressing with sand paper super-fluous, and the shavings have all the appearance and much of the flexibility of fine paper. In many other ways that might be instanced, the improvement in machinery is not less striking, but what has already been given may sufficiently illustrate the general advance.

The sources from which modern wood-working machinery is obtained are various. Notable firms of machinists through-out this country, in America, and on the Continent, are drawn upon, each of whom, although not furnishing complete installations of wood-working machinery, are distinguished for some "special make" of one or other of the machines necessary. In the plentitude of firms whose names suggest themselves in this connection, it may be invidious to single out any for special mention, yet, of firms in this country, Messrs M'Dowall & Sons, of Johnstone, and Messrs T. Robinson & Son, Roch-dale; and of firms in America, Messrs J. A. Fay & Co., of Cincinnati, may be noticed as having furnished many machines which are highly valued in shipyards.

Notwithstanding the recent advancement in this direction, there is still scope for improved wood-working machinery, and for machines to overtake additional work in shipyards. A single, though perhaps not particularly striking, instance may be given. While attempts have been made to supply it, there is not yet, so far as the author knows, a machine for planing decks after the planking has been laid, and the seams caulked and payed. Those acquaint with the laborious and unskilled nature of the work to be done, will readily concede the fitness of applying, if possible, mechanical means to achieve it.

Attention may here be directed to the subject of improvements in shipyard machines and methods of work, directly due to the careful study of results from every-day practice. Workmen themselves have too seldom been instrumental in effecting such improvements, although in many respects the most fitting mediums through which improvements could come. A lingering antipathy to new machinery on the score of its supplanting hand work, and perhaps the want of proper knowledge of scientific principles, have prevented many from taking part in this way. To encourage the exercise of the inventive faculty amongst workmen, as well as to reap personal advantage, Messrs Denny & Brothers instituted in 1880 a scheme of rewards for invention in their establishment, which has been attended with gratifying success, and has since been copied in other quarters. Particulars of this scheme will be given in the following chapter, thus making detailed reference here unnecessary. It may be said briefly, however, that awards ranging from £12 to £3 are paid to workmen who submit inventions, and when any one has been successful in obtaining five awards he receives a premium of £20, and when he has obtained ten awards he is paid a further premium of £25—the premiums increasing by £5 for every additional five awards received. During the time it has been in vogue as many as 200 claims have been entered, over 110 of which have received awards, representing in all the disbursement by the firm of about £500. The majority of the awards made have been concerned with improvements in the joinery departments. Some of the machines there have been modified or altered so as to do twice the quantity of work previously possible, some to do a new class of work, and others to do the same work with greater safety, and with less wear-and-tear.

In several other sections of shipyard work, progress is strikingly evinced. Of these it may suffice to instance the work of transport between one shop and another, and between workshops and building berths, also that of lifting heavy weights either by stationery or locomotive cranes.   Means of effecting such work are now employed in many yards, which, viewed in the light of former things, are truly prodigious.

The increasing propulsive power with which steamships are being fitted necessitates ponderous weights in connection with the engines and boilers.   The means available for lifting such weights have not until within recent years been possessed by private shipbuilders, but have been the property of public bodies, such as Harbour Trusts.   The majority of shipbuilders have still to depend on such outside aid, but within the past few years several large firms—particularly on the Clyde—who have the necessary dock accommodation, have erected in connection with their works enormous "sheer-legs;" the modern equivalent for cranes, which are now somewhat out of fashion for ponderous work. , Some of these are amongst the most powerful ever erected, being capable of lifting 80, 100, and even 120 tons weight.   Such enormous appliances, it may readily be understood, enables the firm possessing them to be independent of extraneous assistance, and to complete in every respect within their own establishments vessels of the largest class.

The means of transporting material in shipyards by systems of railways laid alongside the principal workshops, and traversing the yard in all directions, have been amplified and improved in many yards within recent times.   Connection is made in most instances with sidings from main lines of railway, whereby materials and goods can be at once brought into the yards from whatever part of the kingdom; and in the largest yards special locomotives are constantly employed doing this work.   In well arranged establishments the railway first enters a store-yard, and the material is lifted from the trucks by travelling-crane or other means, and deposited on either side of the railway, plates being set on edge in special racks, from which they can be easily removed by the workmen.   Leaving this, the lines of railway traverse the building yard throughout, and are designed to permit of the material being conveyed without retrocession, but with the necessary stoppages for its being put through the various courses of manipulation, to the vessel in which it is to be used. A recent and very serviceable amplification of the system of railway transport has been fitted in one of the largest Clyde yards which enables material to be conveyed with greatly increased ease and despatch in directions and to situations wholly inaccessible to the main lines of rails. This is the narrow gauge portable system, patented by M. Decauville, of Petit-Bourg, Paris, which consists of short lengths of very light steel rails, permanently riveted to cross sleepers, and with end connections so formed as to make joint while being pressed into contact. Each section, of 4, 6, 8, 12, or 16 feet long, being complete in itself, the tramway can be laid down in any new situation very rapidly. Where divergences of route take place, curves, crossings, and light turntables are supplied, sufficiently strong to carry working loads, and at the same time light enough to be easily handled. Special waggons and trollies are also supplied by the makers, which, combined with the system of portable rails described, not only worthily take the place of, but far excel in handiness and efficiency, the ordinary wheel-barrows of the shipyard.

List of Papers, &c., bearing on modern shipyard machine-tools, appliances, and methods of work, to which readers desiring fuller acquaintance with the *technique* and details of the subject are referred :—

On the Hydraulic Department in the Iron Shipbuilding Department of the Naval Dockyard at Toulon, by M. Marc Berrier-Fontaine : Proceedings Inst. Mech. Engineers, 1878.

On the Application of Hydraulic Pressure to Machine Tools, by Mr Ralph Hart Tweddell : Trans. Inst. Engineers and Shipbuilders, vol. xxiv., 1880-81.

On Machine-Tools and other Labour-Saving Appliances Worked by Hydraulic Pressure, by R. H. Tweddell: Proceedings Inst. Civil Engineers, vol. lxxiii., 1882-83.

Wood-Working Machinery, its Rise, Progress, and Construction, by M. Powis Bale : London, Crosby, Lockwood & Co., 1880.

On Stamping and Welding under the Steam Hammer, by Alex. M'Donnell : Proceedings Inst. Civil Engineers, vol. lxxiii., 1882-83.

On the Decauville Portable Railway, by M. Decauville : Proceedings Inst. Mech. Engineers, 1884.

CHAPTER VI.

DESCRIPTIONS OF SOME NOTABLE SHIPYARDS.

ALTHOUGH in the preceding chapter the main directions in which progress with respect to shipyard appliances and methods of work have been outlined, the record necessarily fails to cover many minor matters which are still essential to an appreciative view of modern shipbuilding. This want cannot better be supplied than by giving detailed descriptions of some representative shipyards and engineering works throughout the principal centres. The establishments which will be selected for notice are amongst the largest in the several districts, and on the whole represent almost all that is advanced in the shipbuilding industry, while to most of them a special interest attaches through the many high-class vessels produced from their stocks for the better-known shipping lines. On such grounds it is hoped the intelligent reader will find the choice of yards—where there was no alternative but to choose— justified and fitting. Three Clyde shipyards, two on the Tyne, one on the Wear, and one at Barrow-in-Furness, will be described. The accounts are written from authoritative information specially supplied, aided and verified by personal knowledge of the works dealt with, and are chiefly concerned with the capability and arrangement of the several yards. Other matters of a more technical nature, such as the comparison of methods of work in the several districts,* are not dealt with. To

---

* For interesting and reliable information on this head, as well as on other matters dealt with in this and the preceding chapter, see Sir E. J. Reed's excellent treatise on " Shipbuilding in Iron and Steel."

some extent this still differs in individual yards, but modern practice is being more assimilated throughout the districts as time goes on. The first establishment dealt with will be:—

MESSRS JOHN ELDER &, CO.'S
SHIPBUILDING AND MARINE ENGINEERING WORKS,
FAIRFIELD, GOVAN, NEAR GLASGOW.

The progress of shipbuilding and marine engineering on the Clyde may be said to include several more or less well-defined periods or stages, and the student of industrial progress must feel bound to connect with these the name of the late John Elder, a distinguished leader in these important industries, and an engineer whose improvements in the marine engine deserve to rank alongside those improvements which James Watt effected in his day. In 1852 Mr Elder joined his friend, Mr Randolph, in an established business, and shortly afterwards made preparations to add marine engineering to the mill-wright and other businesses of the firm. The new firm speedily established itself through a series of improvements, having for their object the reduction of fuel consumption on board steam vessels. In 1860 the firm commenced to build ships, and as shipbuilders and marine engineers they laboured successfully for sixteen years, building during that period 106 vessels, with an aggregate tonnage of 81,326 tons, and constructing 111 sets of marine engines, showing a nominal power of 20,145 horses. At this time the co-partnery contract expired, and Mr. John Elder took over the entire works, carrying them on with great success until his death, which occurred in London in September, 1869, when at the early age of 45 years. After his death the business of the firm was taken up by Mr. John F. Ure, Mr. J. L. K. Jamieson, and Mr. William Pearce, all of whom had previously achieved distinction in shipbuilding and engineering, and the efforts of these gentlemen far exceeded the success of Mr. John Elder's first firm. In 16 years, as above stated, the latter launched 106 vessels of an aggregate tonnage of 81,326 tons, and constructed 111 sets of marine engines of 20,145 nominal horse-power, whereas the new firm launched in nine years 97 vessels of an aggregate tonnage of 192,355 tons, and constructed 90 sets of marine engines of 31,193 nominal horse-power. About six years ago Mr. Ure and Mr. Jamieson retired from the firm, leaving Mr Pearce sole partner, and during these six years the

activity and enterprise formerly characterising the firm have been
worthily sustained, and the firm has kept in the very front rank.
In maintaining this position, and achieving unprecedented results in
the matter of swift steamships, not a little credit is due to Mr A. D.
Bryce-Douglas, an engineer of well-attested skill, who wields the
sceptre of authority in the engineering section.

The works, which are situated on the south bank of the Clyde at
Fairfield, near Govan, occupy an area of about 70 acres, and comprise shipyard, boiler shop, engine works, and tidal basin.   The
disposition of the various workshops is admirable, and as these are
connected with each other by a broad gauge line of rails communicating with all parts of the yard and the terminus of the Govan
railway, the conveyance of raw material in the first instance, its
location in whatever section of the works it may be specially
designed for, and its transmission in the form of finished items of
structure or outfit to the vessels of which it is to form part, are all
accomplished with ease.

Entering by the south-east gate, the visitor proceeds in the direction of the business offices, his first impression probably being one
of wonder at the immense quantities of iron and steel in plates and
bars covering every available piece of ground, as well as the great
quantity of timber of all dimensions stacked and in racks, maturing
for after use.   Arriving at the offices of the firm, the visitor is
probably first ushered into the draughtsmen's rooms, which, as well
as a large reception-room, contain an extensive collection of models
of the vessels that have been constructed by the firm.   In these
apartments a large staff of draughtsmen are employed in the work
of designing new vessels, and making working drawings of ships
already contracted for.

Following the routine of practical operations the visitor is conducted to the moulding loft, which is 320 feet long by 50 feet wide.
Here the drawings of the vessels are put down full size.   The term
"laying off" is applied to the operation of transferring to the mould
loft-floor those designs and general proportions of a ship which have
been drawn on paper, and from which all the preliminary calculations have been made and the form decided.   The lines of the ship
and exact representations of many of the parts of which it is composed are delineated here to their actual or real dimensions, in order
that moulds or skeleton outlines may be made from them for the

guidance of the workmen. These lines, when completed and carefully verified, are afterwards transferred to scrieve boards, from which the frames, floors, &c., are bent. In connection with the moulding loft is a pattern shop, in which the various moulds required in "laying off" are made.

Descending to the iron-work machine shop, which measures about 1000 feet long by 150 feet wide, a scene of great activity meets the eye. Proceeding to that section where the bending blocks are situated, the operation of forming the frames of a vessel may be noticed. The bending blocks are massive iron plates weighing several tons, on which the form of the frame is marked from the scrieve boards. All over the blocks are round holes, closely spaced and equi-distant, in which iron pins are placed to give the form of the frame to be bent. Long bars of angle-iron, properly heated in adjacent furnaces, are brought by the workmen to the blocks, and there the bars are bent round the pins to the form required. The half frame of a ship is thus fashioned to the proper form in little more time than it takes to describe the process. It is now allowed to cool, and it is then returned to the scrieve boards to be set or adjusted with the reverse frame, which with the floor plate go to make the frame in its finished form. While this is going on, the keel blocks are being laid in the usual manner on the building slip, and the keel, stem, and stern-posts are being forged and drilled. The keel is laid, and the frames are then set up in their places, and are kept in position by shores and ribbon pieces. The stem and stern-posts are then set up, and the work now becomes general all over the vessel. The beams previously made are put up, the bulkheads, stringer plates, and keelsons are added in due succession, and the outside shell is being fitted and riveted. Thus the full and perfect form of the vessel is gradually developed, and exhibits one of the most interesting and useful productions of man's labour. In the bending shop alluded to are several large Gorman furnaces, 25 smithy fires for heating angle irons, several sets of plate-bending rolls, five stands of vertical drilling machines with several spindles each, a huge punching machine capable of producing ten rivet holes at each operation, squeezers, boring, planing, countersinking, plate-bending, plate planing, numerous punching and shearing machines, and other appliances. The motive power of this section is supplied by a powerful set of engines lately erected by the firm.

Immediately to the front of this building are the slips, which extend 1,200 feet along the Clyde, and admit of 12 to 14 vessels being proceeded with at one time. While proceeding among the slips hydraulic riveters may be observed at work on several structural features. The attention given to such machines in the preceding chapter makes further notice here unnecessary.

When a steamship leaves the ways she is towed into the firm's tidal dock to receive the boilers and machinery. With the assistance of a pair of 80-ton sheer-legs, Messrs Elder & Co. are able to complete this part of the construction of a vessel with wonderful despatch. In connection with this section is a smithy and small mechanics' shop, which are alongside of the wharf. Space will not permit a description of the smiths' shop, the paint shop, riggers' loft, plumbers' shop, belt-makers' shop, boat-builders' shop, block and pattern-makers' shop, pattern store, general store, &c., about each of which much of interest might be written.

The wood-working department, though stocked with the most approved labour-saving appliances, still affords employment to several hundreds of hands. In the saw mill, which is about 100 feet square, there are several sets of steam saw frames, circular saws, planing machines for operating on deck planks, and other tools, the producing capacity of which is very large. Adjacent to this is the spar shed, where all the spars required on board the vessels are made.

In the joiners' shops are numerous wood-working machines, which are placed advantageously all through this department, comprising planing, morticing, and moulding machines, circular and fret saws, surface planing and jointing machines, general joiners, lathes, and a variety of other tools from the most noted makers of this class of mechanism. The cabinetmaker's shop is a spacious one, and here the finer class of interior fittings are seen in all stages of progress. Nothing in this section seems omitted in the way of mechanical appliances to afford the utmost facility for rapid production and excellence of workmanship.

The marine engineering department of the business is conducted in an imposing pile of buildings about 300 feet square. This immense shop is 50 feet high, and is divided into four bays, or compartments, by three spacious galleries of two floors, each 30 feet wide, and extending the entire length of the building. These galleries serve the double purpose of supporting powerful travelling cranes (two of which are capable of lifting loads of 40 tons, and the other two lesser weights), and providing convenient retreats where boiler-making, copperwork, and other operations are conducted. It is doubtful if a similar collection of ponderous tools is to be found anywhere else in Great Britain. Notable among the heavy tools seen here in operation is one of enormous proportions for planing and trimming armour plates, being capable of smoothing a surface 20 feet by 6 feet.· There are three self-acting screw-cutting lathes, two slotting machines of great power, a universal radial drilling machine, with a radius of 18 feet, capable of boring a hole 4 inches in diameter, through a 9 inch plate in half-an-hour; a turning lathe having a 10-ft. spindle with a diameter of 20-ins.; a planing machine which cuts either horizontally or vertically, and has a traverse of 15 feet by 12 feet; two vertical boring machines, each with a travel of 5 feet; a turning lathe 8½ feet in diameter, with a 34 feet shaft; and a terrible and mysterious-looking machine, with a metallic disc 18 feet in diameter, armed with powerful steel cutters fixed round its circumference,· which takes a shaving of 2½ inches off the mass of iron upon which it is operating. This machine was the invention of the late Mr Elder's father, and is one of the most wonderful tools in existence. Adjoining this engine shop is the forge, which, with its 50 fires, 16 steam hammers, and all the necessary appurtenances to produce forgings with despatch, is an exceedingly busy section of the works. It is 300 feet long and 100 feet wide; and being lofty, excellent ventilation is obtained.

There are three smithies of large dimensions—one being retained for heavy work, and the others for light work. In connection with the engine shop is a pattern shop which, like all the other woodworking departments of the premises, is fully provided with tools having the most modern improvements. The brass foundry is well appointed, and is arranged in two sections—one for light, and the other for heavy work. Manganese bronze propellers, of which the firm make a speciality, are made here in great numbers; the monthly out-put of this department amounts to 45 tons, all of which is used up in the yard, with the exception of a number of propellers which the firm supply to other shipbuilders.

The capabilities of the Fairfield establishment, it may readily be believed, are of the highest order. Scarcely anything need be said in substantiation of this, as the past few years have witnessed the continuous production from its stocks of very many steamships of the highest class, whose names have already become "household words." Of these it may be sufficient to instance the *Arizona*, the *Alaska*, the *Austral*, the *Stirling Castle*, and the *Oregon*. Apart from these, and perhaps no less worthy examples of Fairfield work, vessels of war have been turned out to a goodly extent, as well as vessels for a great variety of trades, but it is for the fast mail and passenger steamships that the establishment is chiefly famed. Its reputation in this respect bids fair to be augmented by the production of the two powerful Cunard steamers already referred to in this work, and which are now nearing completion.

The following tabulated form shows the amount of tonnage built, and the horse-power of engines fitted, by Messrs Elder & Co. during the past fourteen years :—

Years.	Tonnage. Gross.	H.P. Indicated.	Years.	Tonnage. Gross.	H.P. Indicated.
1870	22,795	18,139	1877	7,704	9,550
1871	31,889	29,000	1878	18,247	11,750
1872	24,510	22,450	1879	16,895	15,510
1873	24,829	18,300	1880	32,775	38,024
1874	31,016	16,110	1881	26,575	43,728
1875	17,818	12,040	1882	31,686	41,192
1876	13,533	16,550	1883	40,115	56,995

During ordinarily busy periods the number of operatives employed by Messrs Elder & Co. reaches six thousand. The united earnings of this great army of workmen amount to over £33,000 per month. As a further indication of the stupendousness of the works, it may be mentioned that on board a single vessel—the *Umbria*—as many as 1,200 workmen have been employed at one time. The supervision of affairs in this great establishment is, as may readily be understood, a matter necessitating numerous "heads," "sub-heads," and departments. The general manager in the shipyard is Mr J. W. Shepherd, a naval architect of well-approved ability.

The second of the three Clyde establishments selected for notice, and one in many ways specially noteworthy is :—

MESSRS WILLIAM DENNY & BROTHERS' LEVEN SHIPYARD,
DUMBARTON.

The firm of William Denny & Brothers, Dumbarton, began the business of iron shipbuilding in the year 1844, in a small yard situated on the east bank of the river Leven. To this they subsequently added the "Woodyard" on the opposite side of the river, which had been occupied for a considerable period by William Denny the elder, builder of the "Marjory," "Rob Roy," and many other notable craft, during the infancy of steam navigation. The composition of the firm at the outset comprised William, Alexander, and Peter, sons of the builder of the "Marjory," but it was augmented after a time by the assumption of two other brothers, James and Archibald. The co-partnery some time after again underwent change when the two brothers Alexander and Archibald seceded, and formed small yards of their own. In 1854 the firm sustained an almost irreparable loss in the death of William, the original promoter of the concern, to whose energy and surpassing skill most of the success then attained was due. His decease was deeply lamented, not only as an irreparable family bereavement, but as a public loss. When he first devoted his energies to the formation of an iron shipbuilding concern, it was at a time of great industrial gloom in the community. With its successful establishment began a brighter era in the industrial and social history of the burgh—one which has never once been seriously interrupted, and seems only now to be approaching the "high noon" of its prosperity. Sometime subsequent to the decease of William, the co-partnery was further reduced through the death of James. For a considerable time thereafter the business was carried on by Peter alone, until in 1868 he was joined by his eldest son William, and 1871 by Mr Walter Brock—co-partner in the firm of Denny & Coy.: a distinct marine engineering business established by Peter Denny and others in 1851. Within the past three years further accessions to the firm have been made in Mr James Denny, son of James of the original firm, and in Messrs Peter Denny, John M. Denny, and Archibald Denny, sons of Peter, and younger brothers of William, who for some time has been managing partner of the shipbuilding firm, as Mr. Brock is of the engine works.

In 1867 the firm transferred their establishment to the present site on the east bank of the river Leven near its confluence

with the Clyde, and under the shadow of the Castle-rock, which
figures largely, alike in the scenic renown and the historic annals of
Scotland.   Through a most elaborate series of extensions and
improvements carried out within the past two-and-a-half years, the
works have been enlarged to more than double their previous
dimensions, and correspondingly increased in working capability.
They occupy a total area of forty-three acres, over five acres of which
are taken up with wet dock accommodation, and as much as seven-and-
a-half acres with workshops, sheds, and roofed spaces of various
kinds.   The yard has a most advantageous and extensive frontage
to the Leven, which, under the provisions of a recently obtained
Harbour Act, is being greatly improved as regards width and deepen-
ing.   The principal launching berths, eight in number, are ranged
about the centre portion of the yard's length, and their projections
into the river Leven, favoured by a bend at this part, are almost in
the direct line of its course.   Through the recent improvements,
these berths are capable of receiving vessels of dimensions and tonnage such as the present race for big ships has not even approached.
The arrangement permits of eight vessels being built of lengths
ranging gradually from a maximum of 750 feet downwards.   Besides
these principal berths, there are spaces near the south end of the
yard, where light-draught paddle steamers and the smaller class of
screw vessels are constructed and launched, or taken to pieces and
shipped abroad.   All the work of construction, fitting out, and
putting machinery on board ship, is accomplished within the yard
gates.   Contributing to this result are two tidal docks, one newly
formed, of over four acres in extent, and another of over an acre.
The bottom of the new dock is 26 feet below the level of the yard
and wharfage, affording at high tide 20 feet of water.   In connection
with the dock, powerful sheer-legs are being erected by Messrs Day
& Summers, of Southampton, capable of lifting the enormous
weight of one hundred tons.   Alongside of the smaller dock
are a pair of sheer-legs, capable of lifting 50 tons, with two sub-
sidiary cranes of 10 tons each.   For all purposes, either of construction or outfit of the largest vessel, these and the other enlarged
resources place the firm in a position of entire independence with
regard to extraneous accommodation or appliances.   The engines
and boilers for Messrs Denny Brothers' vessels are invariably supplied by Messrs Denny & Company, whose large works, greatly

extended within recent years, are situated further up the Leven.
Along the eastern boundary of the Leven Shipyard, for over
1000 feet of its length, the joiners' shops, blacksmiths' shops,
machine sheds, outfit stores, &c., are ranged. The joiners' shops
are most admirable for the completeness of their appointment.
They occupy the ground floor and first flat of a three-storey
building, 250 feet by 65 feet, forming part of the range spoken
of. The machines contained in these apartments are of the
newest and most approved description of both British and American
make, and embrace moulding, planing, mortising, tenoning, dove-
tailing, nibbling, scraping, and sand-papering machines; circular,
band, and cross-cut saws; also machines for decorative carving and
incising, &c., the whole being driven by a special engine of considerable power, located near the building. A large saw-mill and
shed, containing various wood-working machines, are situate close
to the Leven, near the south end of the yard, and all the wood
employed in the yard is here cut from the rough. The blacksmiths' and angle smiths' shops and the machine sheds are correspondingly well furnished with the most modern appliances. The
former of these contain over fifty fires, and ten steam-hammers,
as well as verticals, lathes, &c., conveniently situated. The latter are
splendidly equipped, containing several large plate rolls, planing
machines, beam-bending machines, and an assortment of multiple
drills and counter-sinking machines of the most modern type;
also a large number of punching and shearing machines, including two man-hole punches capable of piercing 30 by 20-in. holes
in plates $\frac{3}{4}$-inch thick. The plate and frame furnace, bending
block, and scrive board accommodation throughout the yard, is of
extent commensurate with the other features above described, all
of which being of recent formation, are of the most approved and
modern description.

The system of railways throughout the shipyard is of an unusually
complete description. Connection is made with the main line
of the North British Railway, and enters the yard on its north
side, where a store yard of about two acres affords ample storage
accommodation for material in steel and iron. Leaving this and
traversing the building yard throughout, the lines of railway are
designed to permit of material being conveyed without retrocession
to the vessel of which they are to form part, but with the stoppages

necessary for their being put through the various courses of manipulation.   In addition, the yard is traversed in directions and to
situations inaccessible to the main lines of rails, by the narrow gauge
portable system, patented by M. Decauville, which is of great service.

A special department in the establishment of Messrs Denny, and
an entirely novel feature in a private shipyard, is the experimental
tank, already referred to in the Chapter on scientific progress. This
notable section of Messrs Denny's works may be described as consisting of a basin 300 feet long, 22 feet wide, and containing
9 feet of water over the principal portion of its length.   Around
this basin are the shops and appliances for the work which has to
be done—constructive, experimental, and analytical.   This work on
the constructive side consists of making paraffine models, which
represent on an appropriate scale the ships to which the experiments
have reference ; the paraffine is melted, cast in a rough mould to
the approximate shape, and afterwards faired off by a specially-
constructed and very ingenious cutting machine When finished the
model is passed on to the second stage—the experimental. A stationary
engine draws a carriage along a railway suspended above the water
space, the carriage is accompanied by the model, with an attachment which allows the model to move freely, and at the same time
to depend entirely for its propelling force upon a spring carried by
the carriage.   The extensions of this spring are measured and
recorded automatically, so too are the speeds, the record being made
by electric pens in the form of diagrams, on a revolving cylinder which
is part of the apparatus of the carriage.   The analytical work consists
of obtaining from the diagrams the items of speed and propelling
force, the relation between which, at all speeds for which the
experiments have been made, is thus obtained.   The facilities which
are offered by the tank for investigating to the utmost the laws of
hydrodynamics in so far as they affect, practically, the resistance of
ships, is thus obvious.   On the facade of the tank, fronting the
public street, Messrs Denny have placed an admirably-sculptured
medallion portrait of the late Mr. William Froude, of Torquay,
the noted experimentalist.   Underneath is the following inscrip-
tion :—"This facade of the Leven Shipyard Experimental Tank
is erected in memory of the late William Froude, F.R.S., L.L.D.,
the greatest of experimenters and investigators of hydrodynamics.
Born 29th November, 1811.   Died 14th May, 1879."

Telephonic communication having previously been established with advantage between Leven Shipyard and the Engine Works of Messrs Denny & Co., towards the close of 1883 a telephone exchange system was established in the shipyard, by which means twenty-six separate places are in communication with one another. These are the residences of the principal members of the firm, the manager's house, the Levenbank Foundry, the Dennystown Forge, four stations at the Engine Works, and seventeen stations within the shipyard, representing in all from six to seven miles of line wire. The electric light has already been partially introduced into the shipyard, but steps have been taken by the firm for further extending it to the various offices, the experimental tank, the joiners' shop, and the upholstery and decorators' rooms, as well as providing arc lamps of great power to light up the area of the yard itself.

Besides the introduction of the electric light into their yard, Messrs Denny have formed an electrical department in connection with their works, which will not only be employed in arranging and maintaining the yard installation, but will also undertake the fitting of the electric light installations on board vessels built in the yard. To supervise and manage this important department—which, it may be remarked, is entirely novel as a branch of shipyard work—the firm have engaged the services of a skilled electrician, under whom a staff of operative electricians are employed.

On account of the increased employment it brings to their towns-people, and also doubtless on grounds of increased economy and efficiency, Messrs Denny seek to overtake, as much as possible, the entire work connected with a ship's construction and outfit in their own establishment. Towards the close of 1881 they began the introduction of a department for the designing, decoration, and furnishing of the saloons of their vessels. This department is now of established importance in the yard, and embraces four more or less distinct branches. Firstly, the architectural and decorative designs of the various saloons are determined upon by what may be called the architectural branch, under the immediate supervision of a professionally-trained architect. The work of practically carrying out these designs is at present entrusted to three sections of workers. (1) The decorative department, proper, which overtakes the painting of the various ornamental panels, dados, friezes, &c., of the saloons, and the staining of the coloured glass used in saloon windows, skylights, doors, &c. (2) The carving department, in which the carved work fitted on the bow and stern of vessels, also the numerous small pieces of carved work introduced into the architectural arrangement of the saloons, are overtaken. (3) The upholstery department, in which all the work connected with upholstering the saloons and state-rooms—usually, in other yards, made the subject of sub-contract—is overtaken from first to last. In this branch female labour is employed to a considerable extent, while much of the decorative painting referred to above is also done by females. Under the guidance of a lady artist, the employés in this branch have evinced much aptitude and taste for the work.

Successive enlargements and increased appliances have now rendered the Leven Shipyard capable of turning out from 40,000 to 60,000 tons of shipping per annum. The work hitherto achieved has been almost exclusively that of steamship building, but inside of that general limitation it has been of a varied and comprehensive description. Steamships for many of the largest ocean and coast-trading companies, gunboats and transport ships for foreign Governments, and light-draught paddle-steamers for the rivers Volga, Danube, Ganges, and Irrawaddy, have all been furnished from the stocks of Leven Shipyard. The accompanying list, which is of work done during the period of the firm's existence, viz., since 1844, affords at once an adequate conception of the large amount of important work done for the better known shipping companies:—

	No. of Vessels.	Tonnage.
British India Steam Navigation Co., -	50	107,060
Peninsular and Oriental Steam Navigation Co.,	15	89,171
Austrian Lloyd's Steam Navigation Co.,	16	27,191
J. & A. Allan, Glasgow, Allan Line.	11	24,530
J. & G. Burns, Glasgow,	20	21,101
Union Steamship Co., New Zealand.	19	19,700
A. Lopez & Co., Cadiz, - -	7	19,178
British and Burmese Steam Navigation Co.,	12	18,837
River's Steam Navigation Co , -	18	10,678
Union Steamship Co., Southampton,	2	6,227
Irrawaddy Flotilla Co.,	14	6,006

Adding to this record the work finished since the close of 1883 and presently on hand, the total for the British-India Company is increased to 115,960 tons; that for the Union Company of New Zealand to 21,260, and an addition is made to the list in the two large steamers *Arawa* and *Tainui,* for the Shaw, Savill, & Albion

Company, which together make about 10,000 tons.   The following exhibits in tabular form the number and tonnage of vessels built by the firm from their beginning the business of iron shipbuilding in 1884 up to and including 1883:—

Year.	No. of Vessels.	Tonnage.	Year.	No. of Vessels.	Tonnage.
1845	3	365	1865	6	4,543
1846	3	252	1866	8	10,867
1847	6	1,007	1867	4	9,154
1848	3	618	1868	8	9,855
1849	6	2,173	1869	12	13,227
1850	5	1,577	1870	4	8,852
1851	5	1,460	1871	7	14,922
1852	5	6,622	1872	6	14,056
1853	7	5,163	1873	7	18,415
1854	5	4,380	1874	9	18,475
1855	6	5,443	1875	9	17,191
1856	7	7,436	1876	5	4,394
1857	5	2,822	1877	10	10,533
1858	3	5,29	1878	18	22,054
1859	5	5,903	1879	13	16,138
1860	2	1,897	1880	12	18,114
1861	4	8,463	1881	8	17,455
1862	5	4,271	1882	13	22,010
1863	9	9,745	1883	10	22,240
1864	13	11,239			

The firm, it may be stated, is now engaged in the construction of their 300th vessel.   Notwithstanding the work of re-arrangement and enlargement which has been under progress for two years or more, the work turned out during that period has been in no way behind as compared with other periods—a fact which eloquently testifies to the administrative ability of those in authority, and to the skill and energy of Mr John Ward, the general manager of Messrs Denny's large works.

In August, 1880, the firm issued a notice to their workmen stating that, having observed during the previous two years many improvements in methods of work and appliances introduced by them into the yard, they very readily recognised the advantage accruing to their business from these efforts of their workmen's skill, and were desirous that they should not pass unrewarded.   The notice further stated that to carry out this desire an Awards Com-

mittee had been appointed, which would consider any claims made
by the workmen, and grant an award in proportion to the worth of
the improvement made, the amount in no case to be more than £10,
or less than £2. The committee then appointed, and which still
holds office, was composed of well-known local gentlemen, in
every way competent to adjudicate. Fully a year later the firm
announced that in the case of an invention thought worthy of a
greater award than £10, they had empowered the Committee to grant
such an award, or were willing,.in addition to giving an award of
£10, to take out at their own expense provisional protection at the
Patent Office on behalf of the inventor, so that he might either
dispose of his invention or complete the patent, provided always
they had free use of the thing patented in their own establishment.
From the reports which have yearly been issued by the committee, it
is apparent that considerable success has attended the scheme. The
number of claims made since its institution has been as follows:—
In 1880, 12; in 1881, 32; in 1882, 27; in 1883, 20; in 1884
(till July only), 91 ; total, 182. Awards have been granted as
follows:—In 1880, 5; in 1881, 22; in 1882, 21; in 1883, 18; in
1884 (till July only), 27; total, 93. It is worthy of note that
about one-half of the awards have been gained by workmen in the
joiner's department. Some of their machines have been modified
or altered so as to do twice the quantity of work previously possible,
some to do a new class of work, and others to do the same work
with greater ease and safety. Four inventions have gained the
maximum award of £10, viz., (1) an improvement made on ships'
water-closet and urinal; (2) the invention of a machine to cut
mouldings imitative of wicker work; (3) an improved arrangement
for disengaging steam and hand-steering gear on board ship; (4) an
improved method of laying the Decauville railway across the main
line. In connection with this latter invention, the patentee of the
Decauville railway system supplemented the committee's grant to
the extent of £10. In a note to last year's report, the firm state
that they have decided to increase the maximum grant from £10 to
£12, and the minimum from £2 to £3; and that in the case of two
men being engaged at the same invention, should it be found worthy
of an award, each will receive at least the minimum award of £3.
A still more recent announcement states that "whenever any workman has received as many as *five* awards from the committee, reckon-

ing from the time the scheme came in force, he shall be paid a premium of £20, when he has received as many as *ten* awards he shall be paid a further premium of £25—the premiums always increasing by £5 for every additional five awards received. Already, it may be stated, four separate workmen have received *five* awards, and become the recipients of the £20 premium.

With regard to the employment of females in Messrs Denny's yard, it may be interesting to state further that the total number generally employed throughout the works amounts to between 80 and 100. In addition to the numbers employed in the decorative and upholstery departments, already noticed, a large contingent are engaged in the polishing rooms, and a further number in the drawing offices as tracers. The employment of females as tracers in shipyard drawing offices, it may be stated, is of recent date. The system had previously been in operation at the locomotive works of Messrs Dübs & Co,, and Messrs Neilson & Co., of Glasgow. Having proved a success there, it has been gradually adopted by shipbuilding and engineering firms on the Clyde, and more recently on the Tyne. The staff in Leven Shipyard consists of 20 members, four of whom are employed in the experimental tank department. All the girls are selected by written competitive examination, the subjects of examination being arithmetic, writing to dictation, and block-letter printing. At first it was intended the girls should simply be trained as tracers, but they displayed such aptitude that to tracing was added the inking-in of finished drawings and the reduction of plans from a greater to a less scale. This they do with a very fair degree of accuracy and neatness. The experienced members of the staff are now employed making displacement calculations, including plotting the results to scale, centre of buoyancy, and meta-centre calculations; calculations of ships' surface, working up and plotting of speed trial results, stability calculations. Most of these calculations are made out on prepared printed schedules, and the whole of the work is superintended by a member of the male staff. In the work of calculation the girls, it may be stated, make large use of such instruments as the slide rule, Amsler's planimeter and integrator. To secure clearness and uniformity in the work of writing titles, data, scantling, &c., on the various drawings and tracings, it was found advisable to train the females in the art of lettering these features in a uniform style of lettering in place of writing them.   In this work they display considerable proficiency and expertness, the results being uniformly legible and well arranged.

Before passing from the subject of female employment in Messrs Denny's establishment, attention should be drawn to one fact, of which assurances have been given by those well informed in the matter. In no instance has the employment of females led to the displacement of men as yard operatives.   Those departments into which females have recently been introduced are now numerically as large as before the innovation.   In some cases, indeed, the numbers are greater than before ; new avenues of labour, and greater elaboration of the old, being the grounds of need for the accessions.

The other establishment selected for notice from the Clyde district is:—

MESSRS J. & G. THOMSON'S

SHIPBUILDING AND ENGINEERING WORKS,

CLYDEBANK.

The business of this firm was founded in 1846, by Messrs James & George Thomson, father and uncle respectively of the present members of the firm.   Originally the firm were engineers, but in 1851, shipbuilding operations were commenced, the yard being then situated in the upper reaches of the Clyde.   Twenty years later the increase of the firm's business and the demand for better accommodation for shipping made it necessary for the firm to take new ground.   The present site at Clydebank was therefore chosen for their shipyard, and since its formation many wonderful transformations have been effected.   It is fully twelve years since ground was first broken.   At that time there was neither house nor railway accommodation, and the difficulties were not easily surmountable, and it must have been determined courage and energy that in such a short time not only formed such a large establishment, but created a town, and introduced a railway.   From Clydebank yard, it may be needless to state, many of the most famous vessels of the Cunard, Peninsular, and Oriental and Union Lines have been launched. From its stocks have emanated such well-known vessels as the *Bothnia, Gallia, Thames, Moor, Hammonia,* and the great Cunard liner, *Servia,* while within a very recent period another vessel—the

America—seemingly destined to eclipse the fame of all these other notable craft, has been built and sent to sea.

Until about two years ago, the engineering section of Messrs Thomson's business was conducted at Clydebank Foundry, Finnieston, Glasgow. It was then resolved, however, to centralise the works, and thus save the great expense of fitting out vessels away from the yard, as well as secure the increased facilities offered in the management and controlling of large bodies of workmen. This important undertaking has now been accomplished, and the establishment, as now arranged, is equal in extent and working capability to any other private shipbuilding concern. The entire premises occupy about thirty-five acres of land, and comprise building yard, tidal basin, yard workshops, and engine and boiler works. When in full operation the establishment gives employment to over 4,000 workmen. The yard possesses eight building slips, laid out for the largest class of vessels, and owing to their situation—facing the river Cart, which here joins the Clyde—excellent facilities for the launching of vessels are afforded.

Proceeding to describe the works more in detail, as in the case of a personal visit, the first feature that may be noticed is a handsome block of buildings which stands some distance from the main entrance to the shipyard. These buildings comprise the clerical, managerial, and naval architects' offices; also a spacious apartment in which are located splendidly-executed models, and sections of the hulls, of the vessels which have been built by the firm. Passing through the yard large quantities of the raw material of the modern shipbuilder are observed on railway waggons, and in sheds—including iron and steel plates, bar, T, H, Z, angle, flat, channel, tubular, and other forms of wrought iron. This material is brought into the yard by railway, which forms a siding of the North British system about a quarter of a mile distant.

The iron and steel plates are first manipulated in a large shed open at the sides and ends; and measuring some 500 feet by 150. Here are situated a large number of powerful machine-tools—bending and straightening machines, punching and shearing machines, drilling machines, hydraulic riveting machines and the like. Some are of the largest sizes made, one punching machine being a 33-inch gap tool. Several other machine-tools in this large shed have special features worthy of notice, and one in particular, a flat keel plate bending machine, must be referred to with some detail. The machine in question was made by the Messrs Thomson themselves, and constitutes perhaps the latest application of machinery to ship-building purposes. It is supplied by hydraulic power from the accumulator that works the riveting plant—which is on the Tweddell system—and is composed of a number of arms resting on a horizontal bar. The arms are raised or lowered to suit the different shapes required, by means of a hydraulic ram placed at each end and pressing upon the horizontal bars.

Leaving the machine-tool shed, which, by the way, is amply provided, as indeed are the works generally, with travelling and fixed lifting appliances, and while *en route* for the smiths' shop, are observed several isolated punching and shearing and other machine-tools for special purposes, and driven by self-contained engines or hydraulic power. The smiths' shop is a well-arranged work-shop, 600 feet long by 60 feet wide, and contains 108 smiths' fires, besides three furnaces at each end for heating frames and plates, for bending and other manipulative purposes. This department is well supplied with the mechanical contrivances of the forge, including steam hammers of various capacities graduating from 12 cwt. up to over one ton. There are 16 small jobbing hammers in this shop; a massive 70-cwt. hammer of Messrs Thomson's own make, is used in the production of stern-posts, rudders, and heavy forgings. The smiths' shop is built upon excellent and somewhat unusual prin-ciples, the roof being so constructed as to readily admit of the egress of the smoke from the fires, thus securing good ventilation.

An engineering and machine shop, well equipped with lathes, drills, and other appliances, limited to the operations connected with the production of water-tight doors, steering gears, and the like, is next passed. In close proximity is the riggers loft, where a large staff of workmen, with the aid of mechanical contrivances, mani-pulate the rigging for the vessels nearing completion in the dock. The firm's well-appointed saw mills are provided with a full com-plement of sawing machinery, much of it of a special and very cleverly contrived character. One machine, for instance. is capable of cross-cutting and ripping a log into the required sizes right away, without the usual intermediate manipulation. The arrangements for conveying the timber into position, and for removing it when cut, are very complete, and eminently calculated to ensure rapidity of production. In convenient proximity to the saw mills are the "saw-doctor's" quarters. The old-fashioned practice of sharpening the teeth of the saws by hand-filing is discarded here in favour of a more rapid and effective method of obtaining the requisite amount of sharpness and "set." Emery-wheels are employed and accomplish the process with a great saving of time and labour.

Amongst the other departments with regard to which no details need be given, yet all of which are admirably appointed, are the brass foundry and finishing shops, where the brass castings and fittings are prepared. The joiners', carpenters', and cabinetmakers' shops are an important and extensive branch of the Clydebank premises. The building in which they are located measures 220 feet in length, by 156 feet in width. Here the ordinary ship-joinery work is undertaken, and the tasteful and magnificent furnishings, used in the luxurious equipment of the vessels built in the yard, are produced in great numbers. The joiners' and cabinetmakers' shops are provided with a vast number of ingenious sawing, wood-working, as well as the more ordinary joinery appliances, manufactured for the greater part by Messrs J. M'Dowall & Sons, Johnstone, near Glasgow, and Fay & Son, the well-known American house. It is noteworthy that the belting for driving the multiplicity of machines located in this department is all conducted below the floor: in this way a welcome freedom from obstruction, and comparative immunity from danger, is effected.

A word may be added with regard to the engines and boilers used by the firm for driving their machinery. During the day the most of the machinery is driven from these main engines, the chief of which is a 200 horse-power motor, by Messrs Tangye, of Birmingham; and at night the principal machine tools and several of the workshops derive their requisite motive power from the small self-contained engines, which are attached, or are in close proximity, to them.

The engineering and boilermaking section of the works occupies in all a space of about 12,000 square yards. The boiler shop is a large and lofty galleried workshop, occupying an area of 4,000 square yards. It is splendidly equipped with all the most modern appliances for accurate and heavy work. Attention may specially be drawn to an enormous hydraulic riveter, erected by Messrs Brown Brothers, of Edinburgh. This riveter, which is just undergoing completion, is designed with a 6¼ feet gap, and can close with ease rivets up to $1\frac{3}{4}$ inch diameter. It is rendered necessary owing to the tendency to greatly increase pressure since the introduction of the triple expansion engine. An engine of 100 H.P., having a steam accumulator, gives the necessary power for working this, and advantage has been taken of the extra power to actuate a system of hydraulic hoists, winches and capstans, which are being substituted for the coal-devouring and often dangerous donkey boilers and steam winches, usually in use for this purpose. The hoists will also be . applied to the larger latches in order to save manual labour.

When ready to be placed on board ship, the boilers are run down to the dock by means of a tramway, in the foundations of which as many as 600 tons of slag have been packed. The boilers are then lifted on board and lowered to their proper place by means of massive shear-legs, constructed by Taylor, of Birkenhead, which are capable of lifting the enormous weight of 120 tons, and which have a foundation composed of some 700 tons of cement.

The new engine works comprise erecting, turning, and tool shops, smithy, brass foundry, and depot for laying castings and other goods, also large stores. The whole cover an area of about 8,000 square yards, making, with the 4000 square yards occupied by the boiler shop, a total area of 12,000 square yards. Machinery by the well-known makers, Messrs Shanks, Heatherington, Harvey, and others, of the most modern and powerful description, has been laid down, also overhead travelling cranes, by Taylor, to lift 30 and 40 tons respectively. Railways have been introduced throughout the shops, and a 6-ton crane locomotive lifts and deposits castings where required. In fact, everything that the most modern engineering skill could suggest has been introduced in order to fit the place for turning out not only the largest class of marine engines, but also for the saving of manual labour, and it is expected that 50,000 I.H.P. can be turned out per annum. The entire premises, it should be stated, are illuminated by the electric light, partly on the "Brush" and partly on the "Swan" systems. The vessels on the slips and in the dock are also illuminated by electric light applied in a portable form.

Since having commenced shipbuilding operations, Messrs J. & G. Thomson have placed as many as 200 vessels in the water, representing an aggregate of 300,000 tons, and a gross capital value of about £7,500,000. The position, therefore, that Clydebank yard takes amongst the shipbuilding establishments of the United

Kingdom is certainly in the very front rank. The general manager of the extensive works is Mr. J. P. Wilson, a gentleman of extended experience, who has before held similar posts, but none more onerous and exacting. Amongst other of the responsible officials at Clydebank of whom mention should be made Mr. J. H. Biles, the firm's naval designer, occupies an important position and shares in the credit attaching to successful work.

The three yards selected from the Clyde district have now been described, and their distinctive features enlarged upon. In passing to the notices of the yards from other districts, it may be stated that efforts will be made to avoid repetition in details that are essentially similar. The notices will be of a still more general character than those preceding, the only portions where anything like fullness may occur being those concerned with features which are not embraced in any of the Clyde yards. The most stupendous and comprehensive of the works to be noticed are those of:—

PALMERS SHIPBUILDING AND IRON COMPANY, LIMITED,
JARROW-ON-TYNE.

Palmers Shipbuilding and Iron Company, Limited, have their works at Jarrow-on-Tyne, about four miles from the sea. The works embrace both banks of the river Tyne, cover nearly 100 acres of land, and employ about 7,000 persons. They were first commenced in 1851 by Mr Charles Mark Palmer, the present M.P. for North Durham, distinguished for the active part he has taken and continues to take in merchant shipping legislation. In 1865 the works were made into a limited company, Mr Palmer becoming chair. man. It is a saying in Jarrow, with reference to these gigantic works, that the raw ironstone is taken in at the one end and launched from the other in the form of iron steamships, fitted complete with all their machinery, to carry on a large share of the world's commerce. However much this may appear the exaggerated utterance of native pride, it must be declared to be a literal truth. The works include within themselves the entire range of operations, from the raising and smelting of the ironstone to the complete equipment of iron steam and sailing vessels of all sizes. The ore itself, raised at the ·

rate of 1,000 tons per day, is brought round by sea from the Company's own mines at Port Mulgrave, near Whitby, in Yorkshire, and is lifted from the river wharf at the works up to the railway level, along an inclined plane worked by a stationary engine. Coke and coal come into the works from Marley Hill and other collieries in Durham and Northumberland, by the Pontop and Jarrow Railway. The coke is discharged into a hopper capable of holding about 1,500 tons, from the bottom of which the blast furnace barrows are filled through sliding doors, dispensing with manual labour. The four blast furnaces are 85 feet high, 24 feet diameter at the boshes, and 10 feet in the hearth. They are capable of producing over 2,000 tons of pig per week, of which more than one-half is used in the Company's works. The blast is heated to about 1,500° Fahr. in eight "Whitewell" hot air fire-brick stoves of the newest pattern, and there are eighteen kilns for calcining the Cleveland ironstone. The rolling mill forge comprises eighty puddling furnaces, producing over 1,000 tons of puddled bars weekly; which, again, are rolled into plates and angle bars of the largest and smallest sizes used in the trade. There are two forge engines with 36-inch cylinders, one of 4 feet and the other of 5 feet stroke, each driving a roll train and four pairs of 22-inch rolls. There are two plate mills and ten mill furnaces, producing about 1,200 tons of finished boiler and ship plates weekly. Each mill has two pairs of 24-in. rolls, reversed by clutch and crabs; a bar mill with two pairs of rolls, driven by a 24-inch cylinder, produces 120 tons per week; a fourth mill, with four pairs of rolls, driven by two 30-inch cylinders, with 4 feet stroke, produces about 300 tons per week of plates; also a large angle and bar mill, driven by a single engine, having 36-inch cylinder and 4 feet stroke, capable of rolling the very largest angles used in the trade. There is also a sheet mill in the forge. Attached to the rolling mills are shears, circular saws, punching, and straightening presses, all of the newest patterns.

The adjoining department is that of the engine works, which is on the same gigantic scale, and is capable of finishing about forty pairs of marine engines with their boilers, annually, besides a proportionate share of replace boilers and repairs. The department produces its own iron and brass castings and forgings. In the boiler shop of this department vertical rolls for rolling long boiler shell plates were first used, and may be seen in operation. In the

PORTRAIT
AND
BIOGRAPHICAL NOTE.
——
CHARLES M. PALMER.

CHARLES MARK PALMER, M.P.,

CHAIRMAN OF THE PALMER SHIPBUILDING AND IRON COMPANY;
MEMBER OF THE IRON AND STEEL INSTITUTE, ETC.

BORN at South Shields, on the Tyne, in 1822. Son of Mr George Palmer, who was in early life engaged in Greenland whaling, and was subsequently a merchant and shipowner at Newcastle-on-Tyne. Was trained for a mercantile life, and having completed his education in France, became, at an early age, partner with his father in the firm of Palmer, Beckwith & Co., export merchants, timber merchants, and sawmill owners : a firm since styled Palmer, Hall & Co., and of which he is now the senior. In 1845 assumed partnership with Mr John Bowes, the late Sir William Hutt, and the late Mr Nicholas Wood, in the Marley Hill colliery and coke manufacture, and subsequently acquiring the colleries of Lord Ravensworth & Partners, and of others, the concern known as John Bowes. Esq. & Partners, has become, under Mr Palmer's sole management, one of the largest colliery concerns in the north of England. In 1852, in partnership at first with his elder brother George, commenced iron shipbuilding at Jarrow, in which year they launched the *John Bowes*, notable as the first screw collier. Through gradual extension the works at Jarrow have become the great establishment described in the body of this work. Many vessels of war have been built by Mr Palmer's firm, and it was in the construction of the iron-clad *Terror*, in their works, at the time of the Crimean war, that rolled in place of forged armour plates were first used, the superiority of the change—since universelly recognised—being then experimentally demonstrated at considerable cost by Mr Palmer's firm. Among other enterprises which owe their existence wholly or partially to Mr Palmer may be mentioned the General Iron Screw Collier Company, the Tyne Steam Shipping Company, several of the great lines of Atlantic and Mediterranean steamers, the Bede Metal Company, the Tyne Plate Glass Company, and Insurance Clubs for Steamers. In politics Mr Palmer is a Liberal, and after unsuccessfully contesting his native town in 1868 he was, in 1874, elected M.P. for the northern division of Durham, a seat which he continues to hold. His country residence is at Grinkle Park, in Cleveland, but Parliamentary and other duties necessitate his being much in London, where he has a town house. The interest he has taken in behalf of the English shipowners has lately resulted in his appointment as one of the new English directors of the Suez Canal.

Yours faithfully
Chas. M. Palmer

year 1882-83, June to June, thirty-six pairs of engines, of 7,300 nominal and 39,240 indicated horse-power, were turned out.

The next department, occupying the east end of the Company's works, is that of shipbuilding. The shops of this department are fitted up with all the newest machines for quick and efficient production of work. It contains the largest graving dock on the coast, also a very fine patent slip, fitted with hydraulic hauling gear. The building slips are suitable for every kind of vessel up to 500 feet in length, and are capable, with those in the Howdon branch of the works on the opposite side of the river, of launching 70,000 tons of shipping annually. There are nine building slips at Jarrow, and six at Howdon. In the year 1882-83, June to June, 35 vessels of the aggregate tonnage of 68,000 tons were built and delivered to their owners. For transporting material throughout the works, three steam travelling cranes and eleven locomotive engines are employed. For discharging ore, two fixed and two travelling steam cranes, also two hydraulic cranes, are in use. At the engine works are sheer-legs 100 feet high, capable of lifting 100 tons—used for lifting engines and boilers, and for masting the vessels.

The output of tonnage by Palmers' Company for 1882 and that for 1883 were severally about double the amount turned out by any other one firm in existence for these years. The following statement of the yearly amount of tonnage turned out by the firm since the commencement of iron shipbuilding on the Tyne in 1852, will be interesting, as showing the gradual strides by which the firm have risen from 920 tons thirty years ago to the wonderful return of 61,113 tons in 1883:—

Year.	Ton.	Year.	Ton.	Year.	Ton.	Year.	Ton.
1852, -	920	1860, -	4,653	1868, -	15,842	1876,	8,635
1853, -	3,539	1861, -	4,751	1869, -	11,900	1877, -	16,235
1854, -	7,469	1862, -	21,493	1870, -	26,129	1878,	23,470
1855, -	5,169	1863, -	17,096	1871, -	19,267	1879, -	36,080
1856, -	7,531	1864, -	22,896	1872, -	12,810	1880,	38,117
1857, -	6,816	1865, -	31,111	1873, -	21,017	1881,	50 192
1858, -	7,625	1866, -	18,973	1874, -	25,057	1882, -	60,379
1859, -	11,804	1867, -	16,555	1875, -	15,819	1883, -	61,113

The first screw-steamer built by the firm, namely, the "John Bowes," well known as the pioneer of water ballast steam colliers, is still in existence, has recently had her engines renewed for the third time, and is now busily employed in her customary service, carrying coals from Newcastle to London.

The general manager of the gigantic works is Mr. John Price, formerly one of the surveyors and a leading spirit in the Underwriter's Registry for Iron Vessels. The following are the other responsible officials:—Assistant general manager and manager of rolling mills, Mr. F. W. Stoker; secretary, Mr. Hew Steele; shipyard manager, Mr. A. Adamson; engine works manager, Mr. J. P. Hall; blast furnaces manager, Mr. P. A. Berkeley; blast furnaces assistant manager, Mr. H. T. Allison; mining engineer, Mr. A. S. Palmer.

SIR WM. ARMSTRONG, MITCHELL & CO.'S SHIPBUILDING WORKS, LOW-WALKER AND ELSWICK-ON-TYNE.

The Low Walker yard of this firm was commenced upwards of thirty years ago by Messrs C. Mitchell & Co., who up to 1883 (when they amalgamated with Sir W. G. Armstrong & Co., the notable firm of engineers and artillerists), had built as many as 450 vessels, or an average of 15 vessels per annum, the average tonnage produced during the last ten years being 23,000 tons. The yard is situated about four miles down the Tyne from Newcastle. It consists of about fifteen acres of ground, and has nine launching-berths, but their arrangement is such that at times there have been as many as fourteen vessels on the stocks. The establishment is laid out in a most modern manner. The space occupied by the building slips has a uniform gradient, and, being perfectly flat laterally, gives the greatest facility in the movement of bogies. The yard is served by two complete systems of railways, respectively on the 4 feet 8-in. and 2 feet 3-in. gauge. The former is in connection with a siding from the North Eastern Railway, whereby materials and goods can be brought from all parts of the kingdom, and two locomotives are constantly employed working the trucks into the yard, one of them being of very special construction, on Brown's patent principle, manufactured by Messrs R. & W. Hawthorn, Newcastle. This locomotive is combined with a steam crane, the jib of which acts as a lever with fulcrum, thus dispensing with chains, and which readily swings right round, depositing the plates on edge into racks arranged on either side of the railway, from which they can be taken with great facility by the workmen at the appropriate time.

The yard is divided in two by a building 250 feet long by 50 feet wide, placed at right angles to the river, and which contains plate furnaces, bolt-maker's shop, plumber's shop, rivet store, tool stores, large bending rolls, straightening machine, and manhole punch, on the ground floor; and on the upper storey rigging loft, sail loft, pattern stores, &c. Along the head of the building berths in one half of the yard there is a line of machine shops 400 feet long by 70 feet wide, in one end of which are installed frame furnaces, bending blocks, &c., as also a number of powerful punching machines, planing machines, special machines for angle cutting, and there has recently been added a powerful radial drill, having four moveable arms arranged to drill holes in any part of plate 16 feet by 4 feet without moving it. At the back of this machine shop, and parallel with it, is a smith's shop 180 feet long by 50 feet wide, fitted up complete with steam hammers, &c. For the other half of the yard there is a large building 200 feet long, and of an average width of about 150 feet, which contains furnace, with bending blocks, &c., several heavy punching machines, planing machines, drilling and other machines; one portion about 80 feet by 60 feet being used as a fitting shop, containing powerful lathes, radial, and other drilling machines on the ground floor, and on the upper floor a lighter class of shaping, drilling, and other machines. In this building are also constructed two drying stoves, wherein the exhaust steam from the engine is used for drying timber. At the upper end of this machine shop is another blacksmiths' shop 130 feet long by 50 feet wide, containing steam hammer and drilling machines for special work. A separate building, 80 feet long by 50 feet, is used for the bending and welding of beams, and is so placed that the beams can be lifted direct from barges alongside quay, and laid in position, ready for use.

The smiths', fitters', and other similar shops are all conveniently situated; and as the vessels lie alongside the quay to be finished off after launching, the minimum of expense in this respect is incurred. There are numerous steam cranes of 10 tons and under on the quays for landing such portion of the material as comes by water, and also to lift articles on board the vessels fitting out.

The saw-mills, joiners' shops, mould loft, &c., are situated at the lower end of the yard, and the appliances for handling and converting timber are most complete. The wood-cutting machinery is very extensive, and embraces most of the newest labour-saving machines. The establishment in full work employs 2,500 men, and has turned out as much as 30,000 tons gross register of shipping in a year, including almost every type of vessel for mercantile and war purposes, which latter branch of work will now have a further development since the amalgamation with the eminent gun-making firm of Sir W. G. Armstrong & Co. For this purpose a new yard has been laid out at Elswick, adjoining the Ordnance Works, which will be of the most complete character.

The site of this new yard comprises about 20 acres, and at first only half-a-dozen building berths will be laid out, but as the frontage is about 2,000 feet, the number of these can be augmented as required. The buildings already erected or in progress embrace a brick built shop, 265 feet long by 60 feet wide, standing at the western portion of the ground, and at right angles to the river. This building is in three storeys, the lower portion being intended for general stores, tool and rivet stores, fitting shop, &c.; the second floor will be entirely used as a joiners' shop, and fitted up in the most complete manner with wood-working machinery of every description. The upper floor will be used as a draughting loft and model room. Parallel to this building, and a little distance from it, will be a blacksmith's shop, 150 feet by 50 feet. Adjoining the larger building above described, and at right angles to it, is the office block, 120 feet by 45 feet. Along the head of the launching berths stands a tool shed 420 feet long by 40 feet wide, containing the ordinary punching, planing, drilling, and other shipbuilding machines, all of the newest and most powerful type. Near the centre of the site is a large shed 220 feet long, consisting of four bays, each 50 feet wide, the whole carried on cast-iron columns, which will comprise the plate and angle furnaces, bending blocks, beam shop, angle smiths' shop, plate rolls, large and small, also keel plate bending machine, &c. The yard is served by a complete system of railways, having a siding from the North Eastern Railway Company's system. Material can therefore be brought from all parts of the kingdom and deposited in any part of the premises.

It is almost unnecessary further to give the particulars of this establishment, suffice it to say that it is being laid out on the experience gained up to date in existing shipyards, and will therefore embrace the newest and most important tools in all branches of work. The intention is that it shall be capable of turning out every description of vessel up to the largest iron-clad, and the construction of war vessels of all kinds will be made a speciality, seeing that the Company can send them to sea completely armed and equipped ready for service. Looking to the magnitude of the establishment, it can be regarded as nothing less than an arsenal, which in time of war would be invaluable to the country. The present and prospective importance of this development of the combined firms' business may. be inferred simply from the fact of the services of so high an authority as Mr. W. H. White, late Chief Constructor of the Navy, having been secured as naval adviser and manager.

DEPTFORD SHIPBUILDING YARD AND REPAIRING DOCKS, SUNDERLAND.

These works, established so far back as 1793, but greatly transformed and extended to suit modern requirements, are owned and presided over by Mr James Laing, son of Mr Philip Laing, their founder. The yard consists of two general sections, situated one on each side of the main road leading to the river Wear. One of these, commonly termed the "Woodyard," is where wood shipbuilding was conducted in the early days, but which now of course, in common with the other section, is used exclusively for iron shipbuilding. The entire works, including offices, docks, brass foundry, and other premises, cover an area of about thirty acres.

The yard embraces the various shops and sheds usually pertaining to building operations in iron, such as iron-working sheds, smiths' shop, joiners' shop, upholsterers' shop, blockmakers' shop, &c., all well equipped with machine-tools and appliances, needful in producing vessels for the most important shipping companies. The two general sections of the yard are each worked by one compound surface condensing engine, all machines being driven by belting from main lines of shafting, no independent engines being fitted. Scriveboards, frame furnace, bending blocks, garboard bender, and other machinery are fitted in each section. Gorman's gas furnaces are used for heating the material, and these, though rather troublesome when first fitted, about twelve years ago, after some alterations in the details, now give complete satisfaction, and surpass in efficiency ordinary coal furnaces. The joiners' shop is situated in the woodyard, and the smith's shop in the other section. In the smith's shop a separate engine is provided to drive the blast, so that if it is desired the woodyard can be kept completely going without having the main engine in the other section at work.

The berths of Deptford yard, have been occupied since the commencement of iron shipbuilding there, over thirty years ago, with vessels for home and foreign shipowners, amongst others for such well known companies as the Peninsular and Oriental Company, the Union Company, the Royal Mail Company, the West India and Pacific Company, the Royal Netherlands Company, and the Hamburg and South American Company. In 1882 the *Mexican*, of 4670 tons gross measurement, the largest passenger vessel ever built on the North-East Coast, and one of the finest of the Union Company's fleet of South African mail steamers, was launched from the stocks of Deptford yard. Including the *Mexican*, the following is the list of vessels launched by Mr Laing in the year named :—

Name.	Material.	Owners.	Gross Tons.
S.S. Friary................	Iron,	British,	2307
S.S. Mount Tabor	do.,	do.,	2302
S.S. Mexican........	do.,	do.,	4669
S.S. Rhosina	do.,	do.,	2707
S.S. Govina	do.,	do,,	2221
S.S. Lero..................	do,,	do,,	2224
S.S. Dolcoath............ ..	do.,	do,,	1824
S.S. Ville de Strasbourg	do.,	Foreign,	2372
S.S. Ville de Metz........	do.,	do.,	2375
Total.....			23,004

At present Mr Laing is building his 301st iron vessel, which represents the 460th vessel produced within the Deptford yard since its commencement in 1793. The work presently on hand chiefly consists of average size steam vessels, combining cargo-carrying powers with high-class accommodation for passengers, several being lighted throughout by electricity, and one being constructed of steel, and having engines on the triple expansion principle.

Connected with the shipbuilding yard there are two graving docks of 300 feet and 400 feet in length, one on each side of the river. One of these is situated at the west side of the iron yard parallel to the building berths, and therefore conveniently placed for all kinds of alterations and repairs to vessels. This dock is kept dry by means of pumps which act as circulating pumps for the conden-

PORTRAIT
AND
BIOGRAPHICAL NOTE.
———
JAMES LAING.

JAMES LAING,

EX-PRESIDENT OF THE CHAMBER OF SHIPPING OF THE UNITED KINGDOM; MEMBER OF THE INSTITUTION OF NAVAL ARCHITECTS; OF THE IRON AND STEEL INSTITUTE; AND MEMBER OF COMMITTEE OF LLOYD'S REGISTER.

MR LAING was born at Deptford House, Sunderland, on 11th January, 1823, and is the only son of Mr Philip Laing, who, as early as 1793, in partnership with his brother John, commenced the business of shipbuilding which, nearly a century later, is still carried on, under greatly transformed conditions, by his son. Mr LAING's earliest impressions and associations were connected with what was afterwards to become his life's vocation, his boyhood having been spent in a home contiguous to his father's yard. While a youth, he served as an ordinary workman in the shipyard, and in 1843, his father, on launching the "Cressy," signalised the jubilee of a singularly successful career by handing over to him the care and titles of the business. Mr LAING continued to build wooden vessels until 1853, in which year the "Amity," his first iron ship, was launched. In 1866 he entirely ceased building in wood, and since then has built a very large number of iron vessels for various owners, amongst others for such well-known companies as the Peninsular and Oriental Steam Navigation Company, the Royal Mail Company, the Union Steamship Company of Southampton, etc. In 1883, he built for the last-mentioned company the Mail Steamer "Mexican," of 4669 tons. Besides the shipyard, he is the owner of graving docks connected therewith, as well as extensive copper and brass works, and is principal proprietor of the Ayres Quay Bottle Works, which are capable of turning out 33,000 bottles per day. For upwards of thirty years Mr LAING has served as a member of the River Wear Commission, and as chairman since 1868. For years he has taken a leading position among shipbuilders and shipowners, not only in his own district, but throughout the country. In 1883 he was chosen President of the Chamber of Shipping of the United Kingdom, and as official representative of that interest has performed signal service, both with reference to the Shipping Bill introduced to Parliament by Mr Chamberlain and the recent agreement come to between the shipowners and the Suez Canal Company, of which company he has since been appointed a Director. For twenty years Mr LAING has acted as a member of the Board of Lloyd's Register of Shipping, and at present is Vice-President of the Load-Line Committee, appointed by the Board of Trade for the settlement of a most important and intricate question. In the shipbuilding and other cognate businesses Mr LAING is now ably assisted by his three sons, Philip, Arthur, and James.

James Laing

sers of the yard engines. The pumps used for emptying this dock, as well as the one on the other side of the river, after a vessel has come in, are of the "Pulsometer" type of large size. The capacity of these docks is such that in one year alone the amount of shipping operated upon, either in the way of repairs, alterations, or simple docking, has reached nearly 60,000 tons. A large number of vessels have undergone the important process of lengthening in these docks —a special and very important branch of shipwork in which Mr Laing has been conspicuously successful. The largest undertaking of this kind was the lengthening of the Peninsular and Oriental Company's screw steamer *Poonah* in 1874 from a length of 315 feet to that of 395, or an increase of 80 feet. The work was satisfactorily completed, and the results of the vessel's after-behaviour at sea were communicated, along with an account of the work of lengthening, to the Institution of Naval Architects by Mr. Edwin De Russett, of the Peninsular and Oriental Company, in 1877.

Adjacent to the shipyard are extensive brass and copper works, employing about 300 hands, which, besides supplying all the brass and plumber work required for vessels in the shipyard, undertake similar work for other shipbuilders, also work for the Navy, such as cast gun-metal rams and stern-posts for men-of-war, and brackets for outer-bearings in twin-screws. All sorts of steam and other fittings—Manchester goods—are also here manufactured and dispersed to all parts of the world. Within the same premises are situated the requisite machinery for effecting repairs to the engines and boilers of vessels overhauled in the docks.

At present a large range of new Commercial and Drawing Offices are being erected near the principal entrance to the yard. A new joiner's shop and sawmill will shortly be erected, and other alterations in the internal economy of the shipyard are contemplated. The new range of offices referred to, have a frontage of about 300 feet, and comprise strong room for the preservation of the firm's books, drawings, &c.; model room, 40 feet in length; foremen's room, 40 feet by 30 feet; general office, 42 feet by 41 feet; private offices for Mr Laing & Sons; drawing office, 45 feet by 40 feet; moulding loft, 78 feet by 40 feet; model-making room, &c. An additional and somewhat noteworthy feature in the new buildings will be a large dining hall for the use of those workmen who have their meals brought to them at the yard. Also, a commodious gymnasium for the benefit of the youth in the employ. These are, in addition to the large "British Workman" already in existence, built by Mr Laing for the use of his employés, and for others who care to subscribe. This institution, comprising dining room, game rooms, smoking room and library, is managed by a committee of the employés, and is self-supporting, a contribution of only one half-penny per week being the qualification for membership, admitting subscriber to all the benefits of the institution.

THE WORKS OF THE

BARROW SHIPBUILDING COMPANY (LIMITED).

The Barrow Shipbuilding Company, Limited, was promoted in 1876 by several gentlemen in Barrow connected with the Furness Railway, the Docks, and Steel Works, chief among whom was Mr. Ramsden (now Sir James Ramsden) then managing director of the Furness Railway, Mayor of Barrow, and leading spirit in its development generally. The Duke of Devonshire, the largest proprietor in the district and in the other public works mentioned, became the largest shareholder and the chairman of the new shipbuilding company, which was then formally constituted, with Mr. Robert Duncan, shipbuilder, of Port-Glasgow, as managing director. Mr. Duncan designed the whole arrangement of the works as they now stand, and continued to act as managing director till 1875, when he resigned, and was succeeded by Sir James Ramsden, with Mr. James Humphreys as manager, which position the latter held till 1880, when he was succeeded by Mr. William John, of Lloyd's Register, to whose talent as a naval architect some tribute has been elsewhere passed in this work.

The total area of the plot of land on which these works stand is 58 acres, with two water frontages, each 1050 feet long, one towards Walney Channel, into which the ships are launched, the other towards the docks where the ships are fitted out. The Walney Channel is sufficiently wide to allow of the launching of the largest vessels without risk, and the site is altogether an exceptionably favourable one. The shipbuilding is carried on in that part of the yard adjoining the Walney Channel, being divided from the engine works by a road, under which is a sub-way, which affords the required communication between the two departments.

Entering the shipbuilding department by the main gate-way in this dividing road the visitor finds himself in a large square, formed by substantial buildings; to the left hand on entering, are the offices, and to the right some of the smaller shops. The opposite side of the square is occupied by the machine shed and smiths' shops, whilst on the right-hand side of the square are the frame-bending shed, and on the left the joiners' shop and the sawmill. Passing through the offices upstairs, the visitor enters a very fine drawing-office and model-room, 100 feet by 50 feet, in which an efficient staff of designers are engaged. On the ground floor are the counting-house, officials' rooms, &c., and beyond these the stores for the supply of everything required in building and outfitting ships and machinery. From the stores, or by the outside square, the moulding loft, 250 feet by 50 feet, is reached, of which the joiners' shop is a continuation. This department is 300 feet long by 60 feet wide, and is fitted with every modern appliance in the way of tools to facilitate work. At the back of this shop is an immense room, 600 ft. by 60 ft., occupied by a sawmill, and used also for spar-making, boatbuilding, and rigging. Above these rooms, in continuation of the drawing office and model-room, from which it may be entered, is the cabinet-making department, which necessarily requires a large amount of space in an establishment where passenger and emigrant ships of the largest types are equipped ready for sea. The iron-working machine-shed, 360 ft. by 100 ft., and the frame-bending shed, 300 ft. by 180 ft., follow in order, occupying the whole of one, and most of the other side of the square above described. Both of these sections are fully equipped with the machinery necessary for the rapid manipulation of material. The smiths' shop, 200 ft. by 120 ft., contains one hundred fires and seven steam hammers, the former being blown by a Schiele fan. Attached to the smiths' shop are shops for fitting smith-work and for galvanizing. All these shops and sheds occupy less than one-third of the ground devoted to the shipbuilding department.

Beyond the machine-shop are the slip-ways, twelve in number, where vessels of an aggregate tonnage of 40,000 tons have frequently been seen at one time in various stages of construction. On these slip-ways have been built the well-known mail steamer *City of Rome* and the steam ship *Normandie*, the largest vessel of the French mail service. Here also were built for the Anchor Line the *Anchoria*, the *Devonia*, the *Circassia*, and the *Furnessia* ; for the Ducal Line, the *Duke of Devonshire*, the *Duke of Buccleuch*, the *Duke of Lancaster*, the *Duke of Buckingham*, and the *Duke of Westminster*. From these slip-ways also emanated the *Ganges* and the *Sutlej* for the Peninsular and Oriental Steam Navigation Company as well as the *Eden* and the *Esk* for the Royal Mail Steam Packet Company. For the Isle of Man Steam Packet Company, the *Ben my Chree*, the *Fenella*, and the *Peveril*. For the Société Anomyne de Navigation Belge Americaine, the s.s. *Belgenland* and *Rhyhland*. For the Castle Line, the s.s. *Pembroke Castle*, and for the Sociètè Generale de Transports Maritimes à Vapeur of Marseilles, the s.s. *Navarre* and *Bearn*. Here were also produced the *Kow Shing* for the Indo–China Steam Navigation Company, and the *Takapuna* for the Union Steamship Company of New Zealand, besides many other vessels well known to the mercantile world. For the Admiralty this yard has turned out seven gun-boats, namely, the *Foxhound*, the *Forward*, the *Grappler*, the *Wrangler*, the *Wasp*, the *Banterer*, and the *Espoir*, as well as four torpedo mooring ships.

Leaving the shipbuilding department, the visitor passes through the afore-mentioned sub-way to the engine works, which occupies an area of ground equal to that of the shipyard proper. To the left may be noticed the coppersmith's shop, the brass foundry, and the engineer's smithy. The Foundry has seven ordinary pot furnaces, and one large reverberatory air furnace for castings of the heaviest class. The smithy is well fitted up with hammers suitable for the work. On the opposite side of the ground are two buildings, the one to the left containing the iron foundry and boiler shop. The foundry, 250 ft. by 150 ft., provided with over-head travellers, is capable of turning out the largest castings required for the monster marine engines of the present day. The boiler shop is the same size, and possesses the most modern contrivances for the skilful and economical execution of work, and it contains a complete equipment of hydraulic riveting machines, both fixed and portable, the largest having a gap of 10 feet and a pressure of 90-lbs.

In the space between the boiler shop and the machine shop there are situated a well-arranged furnace for heating, and the vertical rolls for bending the large plates forming the shells of the marine boilers. In the furnace just mentioned the plates are heated while standing on their edge, and as the top of the furnace is level with the ground, they are readily lifted out by a portable crane and deposited on the bed-plate adjoining the vertical rolls. In this vacant space is also situated the water tower for the accumulator for the 100-ton crane, constructed by Sir Wm. G. Armstrong and erected at the side of the Devonshire Dock, where the machinery is placed on board and fixed for new ships.

The engine shop, although 420 ft. long by 100 ft. wide, is scarcely large enough for the pressure of work oftentimes concentrated there. This shop is unsurpassed in the completeness of its fittings and the perfection of its tools. It, like most of the other shops in the establishment, is fitted up with the electric light.

The foregoing descriptive notes of individual yards may fittingly be supplemented by the following table, which shows the number and relative positions of firms throughout all the districts whose total output of tonnage during the year 1883 exceeded 20,000 tons:—

Firm's Name.	District.	Number of Vessels	Gross Tonnage.
1. Palmer Shipbuilding Co.........	Tyne	36	61,113
2. John Elder & Co..................	Clyde	13	40,115
3. Wm. Gray & Co.	Hartlepool	21	37,597
4. Oswald, Mordaunt & Co.........	Southampton	15	33,981
5. Raylton, Dixon & Co.............	Tees	17	31,017
6. Harland & Wolff................	Belfast	13	30,714
7. Russell & Co....................	Clyde	28	30,610
8. Jos. L. Thomson & Sons........	Wear	16	30,520
9. Short Bros.....................	Wear	14	25,531
10. R. Napier & Sons..............	Clyde	6	23,877
11. Armstrong, Mitchell & Co......	Tyne	17	23,584
12. A. Stephen & Sons.............	Clyde	11	23,020
13. James Laing...................	Wear	9	22,877
14. Pearse & Co....................	Tees	9	22,671
15. Wm. Denny & Bros..............	Clyde	10	22,240
16. Richardson, Duck & Co........	Tees	12	21,413
17. Edward Withy & Co.	Hartlepool	12	21,197
18. Swan & Hunter.................	Tyne	15	20,080

CHAPTER VII.

WITH the change from wood to iron shipbuilding, and with the development of propulsion by steam instead of sails, the shipbuilding industry has become localised and concentrated in those districts which, besides possessing the *sine qua non* of ready outlet to the vast ocean, are specially favoured in being the repositories of immense natural wealth in the form of coal and ores. What may now fairly be considered the great centres of shipbuilding are the valleys of the Clyde, Tyne, Wear, and Tees, and also the Thames and Mersey, although these latter rivers have for a considerable number of years been overshadowed as building centres by the immensity of their shipping. In several other districts, of course, shipbuilding is carried on to a considerable extent, and some of these may yet attain much greater importance than they at present possess. Barrow-in-Furness, notwithstanding the remarkable progress of recent years, is still advancing. Belfast occupies a prominent position, not alone because of the large annual output of tonnage, but by reason of the number of high-class ocean steamships which have been, and continue to be, built there. Dundee, Leith, Hull, Southampton, and other places throughout the United Kingdom, are not without claims to recognition on account of the shipbuilding carried on.

The supremacy of one shipbuilding centre over another in the matter of work accomplished, both with regard to its character and its quantity, not infrequently forms the subject of comment in the columns of journals circulating in the districts concerned. The publication, by these journals, at the close of each year, of the returns of new tonnage produced by the various firms, affords an opportunity for vaunting on such matters, and it is, as a rule, taken advantage of by the compilers of the statements, who are usually members of the staff on the journals in question. These statements, through the interesting nature of the statistics they contain, are widely read, and the labour attaching to their preparation must indeed be considerable. The figures are, as a rule, supplied by the shipbuilders themselves, and from a summation of these the compiler draws his conclusions. The accuracy of the returns and the fairness of the comments based upon them, if not always completely satisfactory, are thus seen to be matters for which the compiler is not wholly responsible.

Frequent exception has been taken by correspondents to discrepancies in the tonnages of individual vessels given in these reports, as compared with the tonnages measured by the Board of Trade officials, and entered in their records. Attention was called to this matter at the close of 1883 by a correspondent in *Engineering,* whose assertions were afterwards corroborated in other journals. From a careful checking of the returns made by the Glasgow press of the shipbuilding on the Clyde for the three previous years this correspondent maintained that the aggregate tonnage was overstated to the extent of about 11,000 per year, or over 34,100 tons for the period named. One very gross instance of the misstatement complained of was given by a second correspondent writing to the *Glasgow Herald,* who drew attention, along with the returns of other firms, to that of a firm building the smaller class of vessels, who were stated in the *Herald's* account to have produced 8,300 tons, when by a careful comparison with the actual tonnages of the vessels as recorded in Lloyd's Register, their total output was found to fall short of the figure given by as much as 2,172 tons, equivalent to 35 per cent. of the actual output. In commenting on these discrepancies several obvious considerations suggested themselves to the critics : such as possible misapprehension, caused by the existence of several kinds of " tonnages," and the difficulty of stating accurately the tonnages of vessels recently launched.

It was questioned, however, after all such allowances were made, whether those furnishing the figures could be exonerated from the sin of carelessness, or indeed, of pure falsification with the view of figuring prominently in the list. The accuracy of these criticisms has not in any way been disproved, nor has any satisfactory explanation been offered.

While no attempt will here be made to solve the matter, it has been felt that, in justice to the subject, these charges could not be ignored when presenting statistics which are derived mainly from the sources thus challenged. Indeed, in comparing for the present work the statistics given by various journals—even in journals confined to the same district—innumerable disparities have been met with, and the agreement has only been *en grosse*. Such being the case, it may be asked, could not other and more reliable sources be consulted? The obvious alternative of using the authoritative returns of the Board of Trade, or of Lloyd's Register, at once suggests itself, but objections to this are even more serious than to using the press statistics. The returns issued annually by the Board of Trade only relate to "Merchant Shipping" registered as such, whereas it is well known that in the returns furnished by the shipbuilders all sorts of vessels built by them are included, and that a very considerable tonnage in war vessels and small vessels for military purposes, also in light-draught river craft, both for our own and other countries, is annually turned out from merchant shipyards. The same objections apply to Lloyd's Register Summary, although, strangely enough, the figures there more nearly correspond with the builders' than with the Board of Trade returns, the information given in both cases being the gross tonnage of merchant shipping built and registered in the United Kingdom. Everything considered, the statistics compiled from press returns more accurately represent the work accomplished throughout the districts than those afforded by any of the sources named. In the statistics which follow, therefore, the press returns have been adopted, but to simplify matters for purposes of comparison—the degree of unreliability warranting it—the terminal figures in large quantities have been reduced or increased to hundredths, according as they have chanced to be under or above fifty.

The fluctuations from year to year in the shipbuilding industry of the principal districts over an extended period is exhibited in an interesting manner by the diagram facing page 188, consisting of curves set up on equi-distant ordinates representing years, to the scale shown on the right of the diagram. The figures from which the curves have been constructed will be found to the left of the diagram.*

It is matter of considerable regret to the author that his utmost efforts to obtain statistics for the Tyne over a period corresponding to that for which the Clyde figures are available have not been rewarded with success. Many likely sources have been consulted, and several gentlemen connected with the river and its industries have been appealed to, but without any satisfactory result. No systematic record of shipbuilding output has been kept by anyone officially concerned with the river, although in every other respect its progress has been abundantly and accurately chronicled. It is only so recently as 1878 that the *Newcastle Chronicle* begun the practice of giving, in the systematic and complete manner for which it is now justly noted, the returns of shipbuilding throughout the Kingdom. To this journal the author is indebted for the figures of work done on the Tyne during the years subsequent to 1878. The figures for the Wear have been taken from an article descriptive of that district appearing in the *Shipping World* for June of the present year.

With regard to the Clyde, it is interesting to observe how in the curve the periods of greatest activity, and consequent, output, are recurrent every tenth year. Thus at 1864, 1874, and, at all events, 1883, the curve forms decided crests as compared with the general undulations over the intervening years.

---

* This method of graphically representing tonnage output was applied for the first time by the author to the Clyde district from the figures supplied by the *Glasgow Herald* for each of the years since 1860, and appeared, with much of the descriptive matter now given, in the issue of that journal for March 4th of the present year.

During the seven years from 1846 to 1852 inclusive the number of steam vessels built on the Clyde amounted to 14 with wood hulls, 233 with iron hulls—total, 247, of which 141 were paddle steamers and 106 screw steamers. The tonnage of the wooden steamers amounted to 18,330, and of the iron vessels to 129,270 tons; the horse-power of the engines in the wooden hulls being 6,740, and in the iron hulls 31,590. In 1851, or nearly a decade earlier than the year at which the curve begins, the number of ships produced was 41, with an aggregate tonnage of 25,320. In 1861, a decade later, 81 steamers were built, the tonnage of which amounted to 60,185, and the horse-power of the engines, 12,493. The tonnage for both steamers and ships, however, during that year was 66,800, as shown by the diagram. During the seven years immediately prior to 1862 the extent and progress of shipbuilding on the river were such that 636 vessels, having an aggregate tonnage of 377,000 tons, were launched from the yards of Glasgow, Greenock, and Dumbarton.

With the year just spoken of a first and very considerable rise in the tonnage output set in and continued till the year 1864, in which year it amounted to 178,500 tons. Various causes of an exceptional nature, or at least, causes apart from the natural progress due to the growth of shipping, were at work in bringing about this increase in the output. The most prominent of these was the necessity which arose for filling up the gaps produced by the withdrawal of many swift steamers from the river and coasting trade to meet the requirements of individuals interested in running the blockade of the ports of the Southern States of America. Between Aprils 1862-3 alone, as many as 30 vessels actively connected in some way with the Clyde and coasting service, were sold for that purpose, and the replacement of these vessels went a considerable way in occasioning the briskness. Another and more abiding cause, however, was the demand for vessels for the cotton-carrying trade. This arose chiefly from the blockade of the American ports, causing cotton to come right from the East Indies and China; and in consequence of the longer

TABLE OF YEARLY TONNAGE

YEARS.	CLYDE TON'AGE	TYNE TON'AGE	WEAR TON'AGE
1860	47,800		40,200
1861	66,800		46,800
1862	69,900		56,800
1863	123,300		70,000
1864	178,500		73,000
1865	154,000		73,100
1866	124,600		62,700
1867	108,090		52,700
1868	169,600		70,300
1869	192,300		72,400
1870	180,400		70,100
1871	196,300		81,000
1872	230,300		181,800
1873	232,900		99,400
1874	262,400		88,000
1875	211,800		79,900
1876	174,800		54,100
1877	169,700		87,600
1878	222,300		109,900
1879	174,800	126,800	92,200
1880	248,700	189,800	116,200
1881	341,000	149,100	148,000
1882	391,900	177,200	212,600
1883	419,600	208,400	212,300

TONNAGE DIAGRAM.

CURVES SHOWING THE ANNUAL AGGREGATE TONNAGE OF NEW SHIPPING PRODUCED IN THE PRINCIPAL SHIPBUILDING DISTRICTS SINCE 1860.

voyage many more ships were necessary to carry on the trade. The fact that more than an average number of wrecks had occurred during the two previous winters, together with an increase of the trade between Britain and France as the result of Mr Cobden's commercial treaty, were elements lending impetus to the briskness in the shipbuilding of the time. ·

In 1865 the output of tonnage was lessened considerably through what appears to have been but the natural course of commerce in its reactionary stage. This lessened activity was much aggravated when 1866 was reached, and in that year a serious interruption to the trade was caused by a lock-out of the workmen consequent on a partial strike made to enforce what the employers considered an unreasonable demand on the part of the men. In 1867 the output was as low as 108,000 tons, but thereafter it took an upward tendency, its rise to the previous level being sudden, but thereafter very gradual, and spread over a number of years. The output kept steadily improving each year, outreaching former totals, until in 1874 the curve, or, as it may be called, the output wave, formed a crest of exceptional altitude. For that year the aggregate output reached the unprecedented figure of 262,430 tons, a result which made natural all subsequent references to 1874 as the " big year." The year 1875, although showing an increase in the number of vessels built, yet fell considerably short of 1874 in the matter of tonnage, thus giving to the output curve a decided downward turn. Matters continued to grow worse during 1876, and many of the Clyde firms had painful experiences of " bare poles " until about the beginning of the year 1877, when a slightly improved state of matters set in. Then there was a general desire amongst the workmen for an advance in wages, which ultimately resulted in the great shipwright strike of midsummer, 1877. This strike, it may be remembered, lasted twenty-four weeks, and was one of the most determined struggles which ever took place in this country, both parties having evidently made up their minds to hold out to the last. The strike culminated in the general lock-out of workmen in the autumn of the same year, which, when withdrawn in favour of arbitration as regards the shipwrights, settled down into a keen fight with the iron-workers. The shipwrights' claim was settled by arbitration, the umpire (Lord Moncrieff) deciding in favour of the em-ployers, and the men accordingly resumed work. The iron-workers' dispute was likewise a difficult matter to decide, but ultimately the men resumed work on the understanding that their claim for an advance upon their wages of 10 per cent. would be considered six months subsequently. The struggles were exceedingly costly alike to masters and workmen, one of the results being seen pretty distinctly in the diminished output of tonnage during 1877.

About the spring of 1878 matters had not improved in any very material sense ; and the ironworkers insisting on a settle-ment of their former claim for an advance, were met by the employers with a proposal to increase the working hours from 51 per week, as arranged in 1872, to 54 hours per week, or to reduce the then rate of wages. The men were not unnaturally averse to the increase of working hours, and signified their opposition. Subsequently a reduction in wages of 7½ per cent. was enforced, with the result that the ironworkers came out on strike for a time. Ultimately in the spring of 1879 a return to the 54 hours was made. The prevailing great depression continued well on into the autumn of 1879. In October of that year the shipbuilding industry experienced an unexpected but very welcome revival, and an unusually large amount of work came to the Clyde. The output which in 1879 had fallen to 174,800 tons, now took a sudden and remarkable jump, the figure for 1880 amounting to no less than 248,650 tons, affording ample grounds for the belief that the impetus at the close of 1879 was no mere temporary spurt, but a solid revival. Subsequent experience has more than justified this belief. In 1881 the output reached the aggregate of 341,000 tons, in 1882 it overstepped even this, and the output curve continued in the ascendant until for the year 1883 the stupendous aggregate of 419,600 tons was reached. Following the course which accepted theories regarding industrial activity and depression suggest, and which actual experience in the past exemplifies, the curve of output ought still to be in the ascendant, reaching its maximum in 1884, and thereafter declining. Although the close of the year is still some distance off, there is already ample reason to believe that this will not hold good for 1884. This result is after all only very natural when the most exceptional activity of the past four years, coupled with the present very unhealthy state of the shipping trade, are taken into consideration.

The history of iron shipbuilding on the North-East Coast district does not commence until the year 1840. In March of that year the *John Garrow*, of Liverpool, a vessel of 800 tons burthen, the first iron ship seen in the North-East Coast rivers, arrived at Shields, and caused considerable excitement. A shipbuilding firm at Walker commenced to use the new material almost immediately, and on the 23rd of September, 1842, the iron steamer *Prince Albert* glided from Walker Slipway into the waters of the Tyne.

During the next eight or ten years very little progress was made, the vessels mostly in demand being colliers, in the construction of which no one thought of applying iron. About the year 1850, the carriage of coal by railway began seriously to affect the sale of north country coal in the London market, and it became essential, in the interest of the coal-owners and others, to devise some means of conveying the staple produce of the North Country to London in an expeditious, regular, and, at the same time, economical manner. To accomplish this object, Mr. C. M. Palmer caused an iron screw steamer to be designed in such a manner as to secure the greatest possible capacity, with engines only sufficiently powerful to ensure her making her voyages with regularity. This vessel (the *John Bowes*), the first screw collier, was built to carry 650 tons, and to steam about nine miles an hour. On her first voyage, she was laden with 650 tons of coals in four hours; in forty-eight hours she arrived in London; in twenty-four hours she discharged her cargo; and in forty-eight hours more she was again in the Tyne; so that, in five days, she performed successfully an amount of work that would have taken two average-sized sailing colliers upwards of a month to accomplish. To the success of this experiment may be attributed, in great measure, the subsequent and rapid development of iron ship-building in the Tyne and East Coast district. The district has maintained by far the largest share—almost amounting to a monopoly—in the production of the heavy-carrying, slow-speed type of cargo steamers, of which the *John Bowes* may be said to have been the prototype.

Statistics for the Tyne, as already explained, are not available to any extent until within recent years,* but from a paper on "The Construction of Iron Ships and the Progress of Iron Shipbuilding on the Tyne, Wear, and Tees," written by Mr. C. M. Palmer, and forming part of the work, "The Industrial Resources of the Tyne, Wear, and Tees," published in connection with the British Association's visit to Newcastle in 1863, it appears that the tonnage of iron ships launched from the Tyne during 1862 amounted to 32,175 tons, and during 1863, to 51,236 tons. Comparing this with the output for 1883—twenty years later—it is found that the figures are more than quadrupled, for in that year the output of the Tyne reached as much as 216,600 tons.

In the year following the launch of the *John Bowes*, namely, in 1853, the first iron vessel built on the Wear, was released from its blocks. The Tees followed the example with great energy and considerable success, and on both these rivers trade in iron shipbuilding has been correspondingly developed.

---

* The following fragmentary returns have, through the kindness of a friend engaged in shipbuilding on the Tyne, been forwarded while those sheets were in the press. They have been gathered from occasional records in the local press, supplemented by personal knowledge, but may only be taken as approximate:—

Year.	No. of Vessels.	Tons.	Year.	No. of Vessels.	Tons.
1864	97	49,820	1868	86	45,390
1865	123	77,500	1869	—	—
1866	110	51,800	1870	95	86,420
1867	81	34,080	1871	—	—

What may be described, however, as the opening of the age of iron on the Wear did not begin till the year 1863. During that year 17,720 tons of iron shipping were launched, and from that time the declension of wood shipbuilding, which had long made the Wear a distinguished shipbuilding port in the United Kingdom, proceeded apace. The causes of fluctuation in the trade throughout the subsequent years cannot be traced with any circumstantiality, but the general progress made can be readily gathered from the subjoined tabular record of the number of ships built yearly, with their aggregate and average tonnage. Wood vessels, it may be stated, formed part of the aggregate till the year 1878, when wood dropped out of the arena altogether:—

Year.	No. of Ships.	Gross Tons.	Average Tons.	Year.	No. of Ships.	Gross Tons.	Average Tons.
1860......	112	40.200	359	1872.......	122	131,825	1081
1861.....	126	46,778	371	1873......	95	99,371	1046
1862....	160	56,920	356	1874......	88	88,022	1000
1863......	171	70,040	410	1875......	91	79,904	878
1864......	153	71,987	470	1876......	60	54,041	901
1865......	172	73,134	425	1877......	75	87,578	1168
1866......	145	62,719	432	1878......	85	109,900	1293
1867...:..	128	52,249	408	1879......	65	92,200	1418
1868......	138	70,302	510	1880......	77	116,200	1509
1869......	122	72,420	594	1881......	88	148,000	1681
1870......	103	70,084	681	1882......	123	212,500	1727
1871......	97	81,903	844	1883......	126	212,300	1685

During the years 1871, 1872, and 1873 the output from the Clyde yards averaged 50 per cent. of the total shipping produced throughout the United Kingdom. That high proportion fell for the years 1874, 1875, and 1876 to as low as 37½ per cent. In 1882 the Clyde's contribution to the grand total did not exceed 32½ per cent., so that in one decade the premier shipbuilding centre has fallen from the proud position of producing half the total shipping built within the United Kingdom to that of turning out less than one-third. Mr

William Denny, dealing with this subject in a paper on the " Industries of Scotland," read before the Philosophical Society of Dumbarton, in December, 1878, attributed the then condition of affairs with regard to the tonnage output of the Clyde to the keen competition of the builders on the North-East Coast of England, who managed to produce their favourite type of heavy-carrying, slow-speed steamers at very much less cost than could be done on the Clyde. Their success in this he attributed to four causes—1st, to the enterprise of the small shipowners and the general public on the North-East Coast of England in supplying capital for steamers of this kind ; 2nd, to the great cheapness of iron in that district; 3rd, to the long hours worked, enabling the shipbuilding plant to be more profitably employed, and to the great development of piecework; 4th, to the fact that all the builders being engaged upon work of the same class, the price of which could be measured per ton of deadweight carried, or per ton gross, and per nominal horse-power, they were able easily to compare the efficiency of each other's yard in point of production, and by that means a keen competition was produced amongst each other. On the Clyde the great variety and frequent speciality of the work prevented any such common measure of prices existing. This way of accounting for the altered relative positions of the chief shipbuilding centres was doubtless at that time the correct one, and to a large extent it still holds true. The productiveness of the North-East Coast ports has in no way declined since, notwithstanding that a larger number of the higher class passenger ships which have long been so much a Clyde speciality are now being constructed there. But the number of yards everywhere have increased in a higher ratio than on the Clyde, and consequently the aggregate of new shipping produced annually in the United Kingdom is made up of a greater number of separate contributions. That this is mainly the reason of the present position of the Clyde relatively to the whole United Kingdom is proved by the figures contained in the accompanying table, which show, amongst other things, that the ratio of tonnage produced by each of the principal districts to the total produced by the whole of them, has not very much altered during the past six years, or since Mr Denny spoke on the subject. If anything, indeed, the Clyde shows in this respect an advance over its northern rivals: although the advance of the Wear during the past two years is equally marked.

Table giving the Number and Tonnage of Vessels Built on the Clyde, Tyne, Wear, and Tees, during the Years 1878-83 inclusive; also showing the Average Tonnage of the Vessels and the Ratio which the Tonnage produced in each District bears to the Total Tonnage:

Districts.	1878.				1879.			
	No.	Tons.	Av'rage Ton'ge.	Ratio to Total.	No.	Tons.	Average Ton'ge.	Ratio to Total.
Clyde.....	254	222,300	875	43·5	191	174,800	915	39·8
Tyne......	115	126,300	1096	24·7	130	139,800	1075	32·0
Wear	85	109,900	1293	21·5	65	92,200	1418	21·0
Tees...,..	37	52,500	1419	10·3	25	31,800	1272	7·2
Totals....	491	511,000		100·0	411	438,600		100·0

Districts.	1880.				1881.			
Clyde,....	209	248,700	1189	44·2	261	341,000	1306	47·0
Tyne......	109	149,100	1367	26·5	123	177,200	1440	24·5
Wear.....	77	116,200	1509	20·6	88	148,000	1681	20·4
Tees......	38	48,500	1279	8·7	34	58,600	1723	8·1
Totals....	433	562,500		100·0	506	724,800		100·0

Districts.	1882				1883.			
Clyde.....	297	391,900	1319	44·6	329	419,700	1276	45·1
Tyne......	132	208,400	1578	23·8	159	216,600	1362	23·3
Wear.....	123	212,500	1727	24·2	126	212,300	1685	23·0
Tees	40	65,000	1625	7·4	44	81,800	1859	8·6
Totals....	592	877,800		100·0	658	930,400		100·0

With respect to the progress of shipbuilding in steel, little requires to be added to the general account given in Chapter I. The tonnage annually produced in steel is a constantly-increas-. ing quantity. Hitherto the Clyde has contributed quite three-fourths of the tonnage of steel vessels, owing chiefly to the vigorous way in which certain of the shipbuilders there have adopted the practice, and also to the openness of the local field for the extensive manufacture of the new material. The North-East Coast, however, bids fair, in the immediate future, to become as productive in steel tonnage as the Clyde district. Recently-discovered processes by which the vast stores of Cleveland ironstone may be made profitably available in steel manufacture are working great changes in the way of modifying old and causing the erection of new works.

The extraordinary growth of steel shipbuilding since its commencement in 1878 is well illustrated by the accompanying tables, which are taken from a paper by Mr. W. John, on "Recent Improvements in Iron and Steel Shipbuilding," read at the meetings of the Iron and Steel Institute in May of the present year. The figures relating to steel may be taken, where any divergence occurs, as more authoritative than those occurring in the general account of work in steel in Chapter I. The tables, however, partake of the imperfections already fully alluded to in the present chapter. With regard to them, Mr. John says:—"Unfortunately, neither of these tables show the actual amount of shipping, either steel or iron, built in this country, because there would have to be a small percentage, perhaps between ten and twenty, to be added to those classed at Lloyds on Table I. for unclassed ships, and there would be a certain proportion, which I am unable to ascertain, to be added to the figures on Table II. for ships built for foreign owners in this country, and not entered upon the British register. However, the figures in themselves are sufficiently significant of the enormous growth of steel shipbuilding within the last six years, and it will be seen at once, as I have said before, that steel as a material for shipbuilding has passed entirely out of the experimental stage, and must be judged henceforth by the results of its working in the shipyards, and by the results of the performances of the ships already afloat,

Table I.—Statement showing the Number and Tonnage of Steel and Iron Vessels Classed by Lloyd's Register of British and Foreign Shipping during the Years 87 8 to 18, both inclusive.

Year	Steel				Iron				Total				Percentage			
	Steam		Sailing		Steam		Sailing		Steel		Iron		Steel		Iron	
	No.	Tonnage	No.	Ton'ge	No.	Tonnage	No.	Tonnage	No.	Tonnage	No.	Tonnage	No.	Ton'ge	No.	Ton'ge
1878	7	4,470			329	406,196	106	111,496	7	4,470	435	517,692	1·6	0·85	98·4	99·15
1879	8	14,300	1	1,700	318	436,339	30	34,630	9	16,000	348	470,969	2·52	3·28	97·48	96·72
1880	21	34,031	2	1,342	324	422,622	31	37,372	23	35,373	355	459,994	6·1	7·14	93·9	92·86
1881	20	39,240	3	3,167	401	622,440	51	74,284	23	42,407	452	696,724	4·8	5·74	95·2	94·26
1882	55	113,364	8	12,477	457	742,244	68	108,831	63	125,841	525	851,075	10·7	12·9	89·3	87·1
1883	94	150,725	15	15,703	576	817,584	68	116,190	109	166,428	644	933,774	14·47	15·12	85·53	84·88

Table II.—Statement showing the Number and Tonnage of Steel and Iron Vessels Built in the United Kingdom and Registered therein during the Years 1879 to 1883, both inclusive.

Year	Steel				Iron				Total				Percentage			
	Steam		Sailing		Steam		Sailing		Steel		Iron		Steel		Iron	
	No.	Tonnage	No.	Ton'ge	No.	Tonnage	No.	Tonnage	No.	Tonnage	No.	Tonnage	No.	Ton'ge	No.	Ton'ge
1879	22	19,522	1	1,700	37	428,082	33	35,332	23	21,222	370	463,414	5·83	4·38	94·15	95·62
1880	26	36,493	4	1,671	82	447,889	39	40,015	30	38,164	401	487,404	6·96	7·26	93·04	92·74
1881	34	68,366	3	3,167	41	590,503	50	68,650	67	71,533	461	659,153	7·43	9·79	92·57	90·21
1882	65	115,449	8	12,478	46	672,740	83	112,852	73	127,927	529	785,592	12·14	14·0	87·86	86·0
1883	92	141,552	11	14,193	88	742,292	72	114,698	103	155,745	620	856,990	14·24	15·37	85·76	84·63

both as profit-earning machines for their owners, by their general wear and tear, for their safety against strains at sea, and in cases of collision and stranding."

CHAPTER VIII.

THE PRODUCTION OF LARGE STEAMSHIPS.

APART from the enormous aggregates, no feature of the annual output of new tonnage during recent years has been more remarkable than the great number of full-powered and capacious steamships built for the various ocean-trading companies. The very general interest with which what has been termed "the race for big ships" was regarded two or three years ago has now settled down into the complacent indifference with which matter-of-fact, every-day things are treated. The number of vessels above 4000 tons gross register built during the year 1881 alone was over two-thirds of the whole number produced during the ten years immediately preceding, and was exactly double the number built during the previous five years. From these general facts it may be understood why the constant additions made to the "leviathans of the deep" excite comparatively so little interest, except where matters of dimension or mere bulk are supplemented by questions of exceptional speed or novel construction.

The subjoined table of steamships above 4000 tons gross register presently afloat or being constructed affords information interesting from several such standpoints; and shows in what years the product of big ships has been greatest, as well as what proportion of individual credit falls to the various centres engaged in their production. The vessels are arranged in the order of their tonnages, which in every case available is the gross register tonnage. While most of the information conveyed in the table is such as may be gathered separately from the registries, the form in which it has been compiled, and the fact of the moulded in place of the registered dimensions being

given, makes it valuable for reference. Except in a few instances, where it was impossible to obtain them, the dimensions
of the vessels have been supplied by the respective builders.

Before presenting the table, several of the most noteworthy
features of the information it conveys may be pointed to. The
list comprises no fewer than 138 vessels, 50 of which are constructed of steel. The year 1881 occurs twenty-six times in
the subjoined table, that number of vessels over 4000 tons
having been turned out within the year. As already stated,
this number is over two-thirds the total number for the ten
years immediately preceding 1881, and is exactly double the
number for the preceding five years. The year 1882 occurs
twenty-four times, the year 1883 fifteen times, and the present
year—although, of course, subject to possible additions—
twenty-one times.

The following summary gives the number of vessels of the
"leviathan" order launched in each year since 1858—the year
which witnessed the production of the *Great Eastern*—an
achievement as regards size which has not hitherto been
equalled:—

1858	...	1	1865	...	6	1872	...	3	1879	...	4
1859	...	0	1866	...	0	1873	...	9	1880	...	3
1860	...	0	1867	...	2	1874	...	9	1881	...	26
1861	...	0	1868	...	0	1875	...	3	1882	...	24
1862	...	1	1869	...	0	1876	...	0	1883	..	15
1863	...	3	1870	...	2	1877	...	1	1884		21
1864	...	2	1871	...	1	1878	...	2			

The column giving the districts in which the vessels have
been built, shows—what doubtless is already well recognised
—that the Clyde is supreme in this quantitative aspect of
steamship production. That river occurs seventy-nine times
in the table, a number equivalent to 57 per cent. of the total
of all the centres put together. Barrow follows next in order,
but with the relatively insignificant contribution of twelve—
although it is worthy of note that this contribution is entirely
made up by the vessels of one firm: *i.e.*, the Barrow Shipbuilding Company—the Mersey contributes eleven, the Tyne ten,
and the other districts correspondingly lower numbers.

List of Steamships above 4000 Tons Gross Register presently Afloat (or at one time in existence) or Under Construction, arranged in the order of their tonnage, and showing Builders' Dimensions, Material employed in Construction, Names of Owners and of Builders, Date of Building, and Where Built.

No.	Name of Vessel.	Gross Tonnage	Builders' Dimensions.	Material Employ'd	Owners or Managing Companies.	Builders.	Where Built.	Date of Build.
1	Great Eastern,	18,915	680 by 82½ by 58	Iron	Great Eastern Steamship Coy.	J. Scott Russell & Coy.	Thames	1858
2	City of Rome,	8,141	546 by 52 by 58¼	Iron	Barrow Steamship Coy.	Barrow Shipbuilding Coy.	Barrow	1881
3	Etruria,	7,718	500 by 57 by 40	Steel	Cunard Ship Coy.	John Elder & Co.	Clyde	1884
4	Umbria,	7,718	500 by 57 by 40	Steel	Cunard Steamship Coy.	John Elder & Co.	Clyde	1884
5	Servia,	7,392	515 by 52 by 40¼	Steel	Cunard Steamship Coy.	J. & G. Thomson	Clyde	1881
6	Oregon,	7,375	500 by 54 by 39¾	Steel	S.B. Guion & Coy., Guion Line	John Elder & Co.	Clyde	1883
7	Aurania,	7,269	470 by 57 by 39	Steel	Cunard Ship Coy.	J. & G. Thomson	Clyde	1882
8	Alaska,	6,932	500 by 50 by 39'7"	Iron	Guion & Co., Guion Line	John Elder & Co.	Clyde	1881
9	...,	6,500	432 by 51 by 37¾	Steel	National Steamship Coy.	J. & G. Thomson	Clyde	1884
10	Normandie,	6,062	460 by 50 by 37¼	Iron	Compagnie General Transatlantique	Barrow Shipbuilding Coy.	Barrow	1882
11	...,	5,736	440 by 47 by 35	Steel	Soc. Anon. de. Nav. Belg. Aner.	Laird Brothers	Mersey	1883
12	...,	5,600	430 by 45 by 33¼	Iron	Mississipi and Dominion Coy.	Chas, Connell & Coy.	Clyde	1884
13	City of Chicago,	5,600	430 by 45 by 33½	Iron	Inman Steamship Coy.	Charles Connell & Co.	Clyde	1883
14	Austral,	5,588	455 by 48 by 37	Steel	Orient Steam Navigation Coy.	John Elder & Co.	Clyde	1881
15	Pavonia,	5,588	430 by 46 by 36	Iron	... Steamship Coy.	J. & G. Thomson	Clyde	1882
16	Cephalonia,	5,517	430 by 46 by 36	Iron	Cunard Ship Coy.	Laird Brothers	Mersey	1882
17	Furnessia,	5,495	448 by 44½ by 36¼	Iron	Barrow Steamship Coy.	Barrow Shipbuilding Coy.	Barrow	1880
18	City of Berlin,	5,491	488 by 44 by 36½	Iron	Inman Steamship Coy.	Caird & Co.	Clyde	1875
19	Orient,	5,386	445 by 46 by.36'10"	Iron	Orient Steam Navigation Coy.	John Elder & Co.	Clyde	1879
20	Parisian,	5,359	440 by 46 by 36'2"	Steel	J. & A. Allan, Allan Line	R. Napier & Sons	Clyde	1881
21	Kansas,	5,275	435 by 43½ by 35½	Steel	... Warren & Coy.	Charles Connell & Co.	Clyde	1882

22	Noordland,	[illegible]	400 by 47 by 35	Steel	Soc. Anon. de Nav. Belge. Amer.	Laird Brothers	Mersey	1884
23	Arizona,	5,147	450 by 45′2″ by 37½	Iron	Guion & Co., Guion Line	John Elder & Co.	Clyde	1879
24	Missouri,	5,146	435 by 43½ by 35½	Iron	George Warren & Coy.	Charles Connell & Co.	Clyde	1881
25	Eider,	5,129	430 by 46′10″ by 36¼	Iron	North German Lloyds,	John Elder & Coy.	Clyde	1884
26	Ems,	5,129	430 by 46′10″ by 36½	Iron	North German Lloyds,	John Elder & Coy.	Clyde	1884
27	Fulda,	5,109	430 by 45¾ by 36½	Steel	North German Lloyds	John Elder & Co.	Clyde	1882
28	Werra,	5,109	430 by 45½ by 36¼	Steel	North German Lloyds	John Elder & Co.	Clyde	1882
29	[illegible],	5,085	395 by 44½ by 33¼	Iron	T. R. Oswald	Oswald, Mordaunt & Coy.	S'ampt'n	1883
30	City of Pekin,	5,079	420 by 47 by 38½	Iron	Pacific Mail Steamship Coy.	John Roach & Son	U. States	1874
31	City of Rio,	5,079	420 by 47 by 38½	Iron	Pacific Mail Steamship Coy.	John Roach & Son	U. States	1874
32	City of Yeddo,	5,079	420 by 47 by 38½	Iron	Pacific Mail Steamship Coy.	John Roach & Son	U. States	1874
33	Arawa,	5,026	420 by 46 by 32	Steel	Shaw, Saville & Albion Coy.	Wm. Denny & Brothers	Clyde	1884
34	Tainui,	5,026	420 by 46 by 32	Steel	Shaw, Saville & Albion Coy.	Wm. Denny & Brothers	Clyde	1884
35	Rome,	5,013	430 by 44 by 36	Iron	Peninsular & Oriental S.N. Coy.	Caird & Coy.	Clyde	1881
36	Carthage,	5,013	430 by 44 by 36	Iron	Peninsular & Oriental S.N. Coy.	Caird & Coy.	Clyde	1881
37	Germanic,	5,008	455 by 46 by 34	Iron	Oceanic Steam Navigation Coy.	Harland & Wolff	Belfast	1874
38	Britannic,	5,004	455 by 46 by 34	Iron	Oceanic Steam Navigation Coy.	Harland & Wolff	Belfast	1874
39	Belgravia,	4,976	400 by 44½ by 34¾	Iron	Henderson Brothers—Anchor Line	D. & W. Henderson	Clyde	1873
40	[illegible],	4,935	340 by 55 by 36	Iron	Ind. Rub. & Telegraph Works Coy.	C. Mitchell & Coy.	Tyne	1884
41	Valetta,	4,911	420 by 45 by 37	Steel	Peninsular and Oriental S.N. Coy.	Caird & Coy.	Clyde	1884
42	Massilia,	4,911	420 by 45 by 37	Steel	Peninsular and Oriental S.N. Coy.	Caird & Coy.	Clyde	1884
43	Faraday,	4,908	360 by 52¼ by 36	Iron	Siemens Brothers	Chas. Mitchell & Coy.	Tyne	1874
44	England,	4,898	362 by 42 by 37¼	Iron	National Steam Navigation Coy.	Palmer Shipbuilding Coy.	Tyne	1865
45	Elbe,	4,897	420 by 44¾ by 36½	Iron	North German Lloyd's	John Elder & Coy.	Clyde	1881
46	Servia,	4,841	430 by 43 by 35	Steel	Cunard Steamship Coy.	J. & G. Thomson	Clyde	1880
47	Gallia,	4,809	430 by 44 by 36	Iron	Cunard Steamship Coy.	J. & G. Thomson	Clyde	1879
48	City of Richmond,	4,780	427 by 43 by 36	Iron	Inman Steamship Coy.	Tod & M'Gregor	Clyde	1873
49	City of Chester,	4,770	430 by 44 by 37	Iron	Inman Steamship Coy.	Caird & Coy.	Clyde	1873

No.	Name of Vessel.	Gross Tonnage.	Builders' Dimensions.	Material Employ'd.	Owners or Managing Companies.	Builders.	Where Built.	Date of Build.
50	Paramatta,	4,759	420 by 43 by 37	Steel	Peninsular and ...al S.N. Coy.	Caird & Coy.	Clyde	1882
51	Ionic,	4,753	430 by 45 by 34	Steel	...ic Steam Navigation Coy.	Harland & Wolff	Belfast	1883
52	Ballarat.	4,752	420 by 43 by 37	Steel	Peninsular and ...ntal S.N. Coy.	Caird & Coy.	Clyde	1882
53	...nd,	4,752	440 by 42½ by 31½	Iron	Soc. Anon. de Navig. Belg. ...r.	J. & G. Thomson	Clyde	1867
54	Doric,	4,744	430 by 45 by 34	Steel	...ic Steam Navigation Coy.	Harland & Wolff	Belfast	1883
55	Borderer,	4,740	400 by 44 by 34½	Iron	J ...hn Glynn & Sons	Barrow ...ding Coy.	Barrow	1884
56	Ib ...ria,	4,671	420 by 44½ by 37½	Iron	Pacific Steam Navigation Coy.	J. Elder & Coy.	Clyde	1863
57	Egypt,	4,670	440 by 45 by 38	Iron	National Steam Navigation Coy.	...pl ...ing Coy.	Mersey	1871
58	Mexican,	4,669	380 by 47 by 34	Iron	Uni ...n Steamship Coy.	James Laing	Wear	1882
59	Scotia,	4,667	366 by 47½ by 42.	Iron	Telegraph Conveyance & Main. Coy.	R. Napier & S ...ns	Clyde	1862
60	Liguria,	4,666	420 by 44½ by 37½	Iron	Pacific Steam Navigation ...Gy.	J. Elder & Coy.	Clyde	1874
61	France,	4,648	395 by 44 by 38	Iron	Compagnie General ...que	Cie. Gen. ...tue	S.Nazaire	1865
62	Labrador.	4,612	395 by 44 by 38	Iron	Compagnie General ...e	...ott & Coy.	S.Nazaire	1865
63	...ra,	4,588	420 by 43 by 37½	Iron	National Steam Navigation Coy.	Palmer Brothers & Coy.	Tyne	1864
64	Amerique,	4,584	400 by 44 by 38	Iron	Compagnie General Transatlantique	Scott & Coy.	S.Nazaire	1865
65	Erin,	4,577	420 by 43 by 37½	Iron	National Steam Navigation Coy.	Palmer ...Brs & Coy.	Tyne	1864
66	Styria,	4,557	420 by 42 by 36	Iron	...hard Steamship Coy.	J. & G. Thomson	Clyde	1875
67	Raffaele Rubat-tino,	4,588	400 by 42½ by 32½	Iron	Messageries Gen. Italiana	Palmer & Coy.	Tyne	1882
68	Bothnia,	4,585	420 by 42 by 36	Iron	...rd Steamship Coy.	J. & G. Thomson	Clyde	1874
69	Spain,	4,512	426 by 43 by 36	Iron	National Steam Navigation Coy.	Laird Brothers	Mersey	1881
70	Gha...,	4,499	412 by 44 by 32½	Iron	Messageries Gen. Italiana	Palmer & Coy.	Tyne	1882
71	City of ...M,	4,496	406 by 43½ by 35½	Iron	Inman Steamship ...Gy.	Tod & M'Gregor	Clyde	1872
72	Roman,	4,491	403 by 43½ by 35	Iron	B...tish and North ...ntic Coy.	Laird Brothers	Mersey	1884
73	Tasmania,	4,488	400 by 45 by 34½	Steel	Peninsular and Oriental S.N. Coy.	Caird & Coy.	Clyde	1884

	Name	Tonnage	Dimensions	Material	Owner	Builder	Port	Year
74	Chusan,	4,488	400 by 45 by 31¼	Steel	Peninsular and Oriental S.N Coy.	Earles' Ship. & Eng. Coy.	Hull	1884
75	St. Ronans,	4,484	402 by 42¾ by 35¼	Iron	Rankin, Gilmour & Coy.		Hull	1884
76	Kaikoura,	4,474	420 by 45¾ by 35'4"	Steel	New Zealand Shipping Coy.	John Elder & Coy.	Clyde	1882
77	Kimutaka,	4,474	420 by 44½ by 33'4"	Steel	New Zealand Shipping Coy.	John Elder & Coy.	Clyde	1884
.78	The Queen,	4,457	382 by 42½ by 37	Iron	National Steam Navigation Coy.	Laird Brothers	Mersey	1865
79	Coptic,	4,448	430 by 43 by 33	Steel	Oceanic Steam Navigation Coy.	Harland & Wolff.	Belfast	1881
80	Stirling Castle,	4,423	420 by 49½ by 32¾	Iron	Thomas Skinner & Coy.	J. Elder & Coy.	Clyde	1882
81	Norseman,	4,386	391 by 43½ by 35	Iron	British & North Atlantic S.N. Coy.	Laird Brothers	Mersey	1882
82	Sardinian,	4,376	400 by 43 by 36	Iron	J. & A. Allan	Steele & Coy.	Clyde	1874
83	Arabic,	4,368	430 by 43 by 33	Steel	Oceanic Steam Navigation Coy.	Harland & Wolff	Belfast	1881
84	Grecian Monarch	4,364	380 by 42¼ by 36	Iron	Royal Exchange Shipping Coy.	Earles Ship. & Eng. Coy.	Hull	1882
85	Tartar,	4,339	392 by 47 by 35½	Iron	Union Steamship Coy.	Aitken & Mansell	Clyde	1883
86	Iowa,	4,329	380 by 45 by 35	Iron	George Warren & Coy.	R. & J. Evans & Coy.	Mersey	1879
87	Greece,	4,310	392 by 43 by 37	Ir n	National Steam Navigation Coy.	Palmer' Brothers & Coy.	Tyne	1863
88	France,	4,281	386 by 43 by 38	Iron	National Steam Navigation Coy.	T. Royden & Sons	Mersey	1867
89	Roslin Castle,	4,280	380 by 48 by 33	Iron	Donald Currie & Coy.	Barclay, Curle, & Coy.	Clyde	1883
90	Canada,	4,276	392 by 43 by 37	Iron	National steam Navigation Coy.	Palmer Brothers & Coy.	Tyne	1863
91	Circassia,	4,272	400 by 42 by 34½	Iron	Barrow Steamship Coy.	Barrow Shipbuilding Coy.	Barrow	1878
92	Devonia,	4,270	400 by 42 by 34½	Iron	Barrow Steamship Coy.	Barrow Shipbuilding Coy.	Barrow	1877
93	Isla de Luzen,	4,252	393 by 44½ by 32	Iron	Cie. Gen. de Tobaccs de Filipinas	Oswald, Mordaunt & Co.	S'ampt n	1882
94	Hammonia,	4,247	375 by 45 by 34'2"	Steel	Hamburg American S.P. Coy.	J. & G. Thomson	Clyde	1882
95	Hawarden Castle,	4,241	380 by 48 by 32'10"	Iron	Donald Currie & Coy.	John Elder & Coy.	Clyde	1883
96	Norham Castle,	1,241	380 by 48 by 32'10"	Iron	Donald Currie & Coy.	John Elder & Coy.	Clyde	1883
97	Richmond Hill,	4,225	420 by 47 by 28	Steel	W. H. Nott & Coy.	H. May & Coy.	Clyde	1882
98	Potosi,	4,219	411 by 43 by 35¼	Iron	Pacific Steam Navigation Coy.	John Elder & Coy.	Clyde	1873
99	Ganges,	4,196	390 by 42 by 34¼	Steel	Peninsular and Oriental S.N. Coy.	Barrow Shipbuilding Coy.	Barrow	1881
100	Sutlej,	4,194	390 by 42 by 34¼	Steel	Peninsular and Oriental S.N. Coy.	Barrow Shipbuilding Coy.	Barrow	1881
101	Shannon,	4,189	400 by 43 by 34¼	Steel	Peninsular and Oriental S.N. Coy.	Harland & Wolff	Belfast	1881

No.	Name of Vessel.	Gross Tonnage.	Builders' Dimensions.	Material Employ'd.	Owners or Managing Companies.	Builders.	Where Built	Date of Build.
102	ChateauMargaux,	4,176	385 by 42 by 33	Iron	Cie. Bordelaise de Nav. à Vap.	Chant. de la Gironde	Bordeaux	1884
103	Chateau Yquan,	4,176	385 by 42 by 33	Iron	Cie. Bordelaise de Nav. à Vap.	Chant. de la Gironde	Bordeaux,	1884
104	Italy,	4,169	389 by 42 by 38'2"	Iron	National Steam Navigation Coy.	John Elder & Coy.	Clyde	1870
105	Anchoria,	4,168	408 by 40 by 35¼	Iron	Barrow Steamship Coy.	Barrow Shipbuilding Coy.	Barrow	1875
106	Sydney,	4,166	420 by 43 by 34	Iron	Messageries Maritimes	Messageries Maritimes	La Ciotat	1882
107	Tongariro,	4,163	380 by 45¾ by 33'4"	Steel	New Zealand Steam Shipping Coy.	John Elder & Coy.	Clyde	1883
108	Aorangi,	4,163	380 by 45¾ by 33'4"	Steel	New Zealand Steam Shipping Coy.	John Elder & Coy.	Clyde	1883
109	Ruapehu,	4,163	380 by 45¾ by 33'4"	Steel	New Zealand Steam Shipping Coy.	John Elder & Coy.	Clyde	1883
110	Ludgate Hill,	4,162	420 by 45 by 28	Steel	W. H. Nott & Coy.	Doxie & Coy.	Clyde	1881
111	John Elder,	4,152	370 by 41 by 36¾	Iron	Pacific Steam Navigation Coy.	John Elder & Coy.	Clyde	1870
112	Isla de Mao,	4,141	376 by 42 by 35¼	Iron	Cie. Gen. de Tobacas de Filipinas	Barrow Shipbuilding Coy.	Barrow	1881
113	Navarre,	4,137	400 by 40 by 33¼	Iron	Soc. Gen. de Trans. Marit. à Vapeur	Barrow Shipbuilding Coy.	Barrow	1881
114	Venetian,	4,136	423 by 41 by 31'10"	Iron	Fred. Leyland & Coy.	Palmer & Coy.	Tyne	1882
115	Bearn,	4,134	400 by 40 by 33¼	Iron	Soc. Gen. de Trans. Marit. à Vapeur	Barrow Shipbuilding Coy.	Barrow	1881
116	Mexico,	4,133	400 by 43¼ by 32½	Steel	Campania Transatlantica Mexicana	R. Napier & Sons	Clyde	1884
117	Tamaulipas,	1,133	400 by 43¼ by 32¼	Steel	Campania Transatlantica Mexicana	R. Napier & Sons	Clyde	1883
118	Oaxaca,	4,133	400 by 43¼ by 32¼	Steel	Campania Transatlantica Mexicana	R. Napier & Sons	Clyde	1883
119	Brittania,	4,129	399 by 43 by 34	Iron	Pacific Steam Navigation Coy.	Laird Brothers	Mersey	1873
120	Clyde,	4,124	390 by 42 by 34	Steel	Peninsular and Oriental S.N. Coy.	Wm. Denny & Brothers	Clyde	1881
121	Aconcagua,	4,112	391 by 41 by 36¾	Iron	Pacific Steam Navigation Coy.	John Elder & Coy.	Clyde	1872
122	Goorkha,	4,104	390 by 42 by 31	Steel	British India Steam Navigation Coy.	W. Denny & Brothers	Clyde	1882
123	Thames,	4,101	390 by 42 by 35	Steel	Peninsular and Oriental S.N. Coy.	J. & G. Thomson	Clyde	1881
124	Werneth Hall,	4,100	400 by 43 by 31	Steel	Sun Shipping Coy.	Charles Connell & Coy.	Clyde	1882
125	Virginian,	4,081	422 by 41 by 31'10"	Iron	Fred. Leyland & Coy.	Palmer & Coy.	Tyne	1881

Ship	Dimensions	Company	Builder	Location
India,	390 by 42 by 21	British India Steam Navigation Coy.	Wm. Denny & Brothers	Ryde
Sorato,	390 by 42½ by 35½	Pacific Steam Navigation Coy.	John Elder & Coy.	Ryde
...,	350 by 43 by 36	Compagnie Gital lantique	..., Gen. Tr que	S.Nazaire
India,	400 by 40 by 33	Barrow Steamship Coy.	Robert Duncan & Coy.	Ryde
Merton Hall,	400 by 42 by 30	Sun Shipping Coy.	Gourlay Bari & Coy.	Dundee
Lake Huron,	385 by 42½ by 31½	...a Shipping Coy.	Lon. & Glas. E. & I. Ship. Coy.	Ryde
Cotopasi,	390 by 42½ by 35½	Pacific Steam Navigation Coy.	Iahn Elder & Coy.	Ryde
Kaiser-i-Hind,	400 by 42 by 34	Peninsular and Oriental S. N. Coy.	Caird & Coy.	Ryde
Illimania,	390 by 42½ by 35½	Pacific Sun Navigation Gy.	Ihie & Coy.	Ryde
Tower Hill,	420 by 45 by 28	W. H. Nott & Coy.	John Elder & C yo	Clyde
R. wa,	390 by 43 by 29	British India Associati o	A. & J. Inglis	Ryde
Buenos Ayrean,	385 by 42 by 34½	J. & A. Al hn	W. D nny & Brothers	Ryde
Ethiopia,	400 by 40 by 33	Barrow Steamship Coy.	A. Stephen & Son	Ryde

APPENDIX.

CALCULATING INSTRUMENTS.

The instruments to which references are made in Chapter IV. as having come into use in some of our leading mercantile shipyards by which the calculations undertaken there are rendered greatly more simple, and are more expeditiously made, seem not to be generally known amongst shipbuilders, and as they undoubtedly save much of the labour and time of calculation, without any sacrifice of accuracy, illustrations of them are here given, together with brief notes of their construction and use. For anything, however, like a satisfactory account of the mathematical principles on which these several instruments are based, readers must consult the authoritative sources to which references will be made.

Assuming that the reader appreciates the advantages of shortened calculation, due to the slide rule, or the use of logarithms, the first instrument that may be noticed is one embodying an application of the principle of the slide rule in a remarkably handy and compact form. This is the calculating slide rule invented by Professor Fuller, of Queen's College, Belfast, equivalent to a straight slide rule 83 feet 4 inches long, or a circular rule 13 feet 3 inches in diameter. From the illustration given it may be seen that the rule consists of a cylinder which can be moved up and down upon, and turned round, an axis, which is held by a handle. Upon this cylinder is wound spirally a single logarithmic scale. Fixed to the handle of the instrument is an index. Two other indices, whose distance apart is the axial length of the complete spiral, are fixed to an inner cylinder, which slides in like a telescope tube, and thus enables the operator to place these indices in any required position relative to the outer cylinder containing the logarithmic scale. Two stops—one on the fixed and the other on the outer or movable cylinder—are so placed that when they are brought in contact the index points to the commencement of the scale.

Regarding the manner of using the instrument a few general notes may be given. As in the ordinary slide rule the operations of multiplication and division are performed by the addition

FIG. 24. or subtraction of the parts of the scale that represent in length the logarithm of the numbers involved in the operations.

For example, suppose the following calculation is to be worked out

$$\frac{6248 \times 5936 \times 4217}{7963 \times 4851} = 4049$$

To do this in the ordinary way would keep the smartest arithmetician busy for a considerable time, whereas by means of the instrument under notice the result is attained in little over one minute's time. The motions in the operation are as follows :—Hold the rule by the handle in one hand and move the scale cylinder by the other until the number 6248 is opposite the index attached to the handle portion. Now, move the inner cylinder (by the top) until one or other of the indices (according to the distance of the number from the bottom of the instrument) on the index arm is opposite the number 7963. The scale cylinder is again moved till the number

FULLER'S RULE. 5936 is opposite one of the indices just referred to, and the inner cylinder carrying the index arm is then moved till one or other of the indices is opposite 4851. Finally, the scale cylinder is moved till the number 4217 is opposite one of the indices on the arm; and the result of the whole operation—4049—is found opposite the index first mentioned, *i.e.*, that attached to the handle portion of the instrument.

It may be further explained that the sliding of the scale cylinder until the new number is opposite the index point really involves two operations : one sliding it till the end of the scale is opposite the index point—which subtracts the logarithm of the divisor; and the other sliding it till the next multiplier is opposite the index point—which adds its logarithm to the previous result. Hence, when the operations end with division the scale cylinder must be moved till the end of the scale is opposite the index point.

The second scientific instrument to be noticed is the Polar Planimeter, invented by M. J. Amsler-Laffon, Schaffhausen, Switzerland, the object of which is to find the area of any figure by simply tracing the outline with a pointer, the instrument—of which the pointer is a part—doing all the rest; the

FIG. 25.

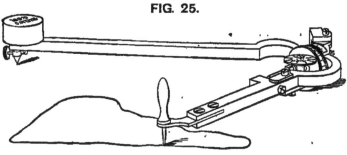

AMSLER'S POLAR PLANIMETER—(FIXED SCALE).

results read off from it having to undergo only a very simple and elementary calculation to attain the desired result.

Planimeters are made of several forms, the two kinds illustrated by Figs. 25 and 26 being the most usual.* The planimeter shown by Fig. 25 represents the instrument as made to one scale only, for square inches of actual measurement. By

---

* These instruments, and the others here noticed, are supplied in this country by Mr. W. F. Stanley, the noted scientific instrument maker of Great Turnstile, Holborn, London. They are described in his treatise on "Mathematical Drawing Instruments," from which work, it should be stated, some of the present notes concerning them are derived. A source of accurate information on the theory of planimeter, to which Mr. Stanley himself expresses indebtedness, is the paper by Mr.—now Sir—F. J. Bramwell, read before the British Association in 1872, and contained in the Association Reports for that year.

its means the areas of, say, cross sections of ship's hull can be ascertained in an extremely short time and with almost perfect accuracy, the readings taken from the instrument having simply to be multiplied by a multiplier consisting of the square of the number of units to the inch, corresponding to the scale on which the sections are drawn, as 4 for ½-inch scale, 16 for ¼-inch, 64 for ⅛-inch, etc.

The Planimeter shown by Fig. 26 is the instrument in a form adaptable to various scales, but does not possess any very marked advantages over the simpler form for the purposes of the naval architect or marine engineer, so that notice of it must be brief. In this form of the instrument the unit can be changed by altering the length of the arm which carries the tracer to

FIG. 26.

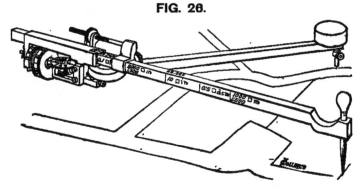

AMSLER'S PLANIMETER—(VARIOUS SCALES).

any of the scales for which the instrument may be made available, and which are found divided upon the variable arm. The scales which are usually provided for are as follows :—

10 sq. in.	= 10 square inches	
0·1 sq. f.	= 0·1 square foot	
1 sq. dcm.	= one square decimetre	
0·5 sq. dcm.	= 0·5 square decimetre	Every total rotation of the roller.
2000 sq. m. } 1: 500	= 2000 square metres on a scale 1: 500	
1000 sq. m. } 1: 500	= 1000 square metres scale 1: 500	

Describing the simple planimeter more in detail, and referring to Fig. 25, it may be said the outline of the figure to be dealt with is travelled round by a pointer attached to a bar moving on a vertical axis carried by another bar, which latter turns on a needle point slightly pressed into the drawing surface. The bar with the pointer is provided with a revolving drum having a graduated circumference and a disc counting its revolutions. The drum is divided into 100 parts, reading into a vernier, which gives the reading of the drum's revolution to the $\frac{1}{1000}$ part of its circumference. Upon the same axis as the drum an endless screw is cut, working into a worm wheel of ten teeth connected with the counting disc, which records the revolutions of the drum.

To use the planimeter, place the instrument upon the paper so that the tracing point, roller, and needle point, all touch the surface at any convenient position. Press the needle point down gently, so that it just enters the paper, and place the small weight supplied with the instrument over it. Make a mark at any part of the outline of the figure to be computed, and set the tracing point to it. Before commencing read off the counting wheel and the index roller. Suppose the counting wheel marks 2, the roller index 91, and the vernier 5, then, the unit in this case being 10 sq. ins., write this down 29·15 (for the proportional or variable scale planimeter this reading would be 2·915.) Follow with the tracing point exactly the outline of the figure to be measured in the direction of the movement of the hands of a watch, until you arrive at the starting point; now read the instrument. Suppose this reading to be 47·67, then by deducting the first reading (29·15) the remainder (18·52) indicates that the measured area contains 18·52 units—*i.e.*, square inches— which is the final result, so far as the instrument is concerned. To obtain the actual area in feet, however, this result must be multiplied by the number before explained corresponding to the scale on which the figure that has been measured is drawn.*

---

* The following is a list of the multipliers for converting the planimeter read ings to square feet for any required scale :—

$\frac{1}{16}$-in. scale	=	256	$\frac{5}{16}$-in. scale	=	10·24	1-in. scale	=	1·00
$\frac{1}{8}$-in. do.	=	64	$\frac{3}{8}$-in. do.	=	7·11	1½-in. do.	=	·44
$\frac{3}{16}$-in. do.	=	28·44	½-in. do.	=	4·00	3-in. do.	=	·111
¼-in. do.	=	16	¾-in. do.	=	1·77			

Assuming the scale to have been ¼-inch per foot, then 18˙52 inches multiplied by 16—the appropriate multiplier for that scale—gives 296˙32 square feet, the exact area.

Several important points remain to be noticed in connection with the use of the instrument. As a rule, the areas to be measured in connection with ship designing are on a small scale, and the fixed or needle point about which the instrument moves can always be placed *outside* the figure measured, in which case the process remains as above stated. It should be mentioned, however, that by placing the needle point *inside* the figure, in such a position as to enable the operator to follow its contour a larger figure can be measured at one operation —the reading, however, being less than the true area by a constant number which varies slightly with the construction of each instrument, and which is found engraved on the small weight already referred to (on the top of the bar in the proportional planimeter). Adding this constant number to any reading taken by the instrument placed as described, gives the true area.

The counting disc may go through more than one revolution forwards or backwards. If the needle point be *outside* the figure traversed the counting disc can only move *forwards* (as 9, 0, 1, 2, &c.) : that is, provided the figure has been traced in the manner directed—in the direction of the hands of a watch. Then as many times as the zero mark passes the index line add 10˙000 to the *second* reading. If the needle point be *inside* the figure, the disc can move either forwards or backwards. If moving backwards, as 2, 1, 0, 9, &c., then add 10˙000 to the *first* reading.

Before passing from the subject of the planimeter it may be both interesting and useful to give an example of a calculation involving its use. Subjoined is a specimen displacement and longitudinal centre of buoyancy calculation, and any one familiar with the prodigious array of columns and figures per. taining to a "displacement sheet" of the ordinary kind cannot fail to appreciate the advantages of the specimen, both with respect to simplicity of arrangement and curtailment of the amount of calculation ordinarily involved :—

EXAMPLE OF SHIP DISPLACEMENT, WORKED OUT BY PLANIMETER.

No. of Sections for Displacement	Area of Half Sections		Simpson's Multipliers.	Functions.	Multipliers for Centre of Buoyancy.	Moments for Centre of Buoyancy.
	Successive Readings of Planimeter.	Difference between Readings = Area in sq. ins.				
1	52·73	0·0	1	0·0	0	0·00
2	52·73	1·82	4	7·28	1	7·28
3	54·55	4·43	2	8 86	2	17·72
4	58·98	5·63	4	22·25	3	67·56
5	64·61	6·12	2	12·24	4	48·96
6	70·73	6·32	4	25·28	5	126·40
7	77·05	6 32	2	12·64	6	75·84
8	83·37	6·27	4	25 08	7	175·56
9	89·64	6·11	2	12·22	8	97·76
10	95·75	5·7	4	22·8	9	205·20
11	01·45	4·64	2	9·28	10	92·80
12	06·09	2·48	4	9·92	11	109·12
13	08·57	0·0	1	0·0	12	0·00
	08·57					

```
(Com· int.)  (mult. for ¼th scale) (both sides)  ⎫  168·12   168·12 ⎞ 1024·20
   28·6    ×       16        ×      2           ⎬  =8·716          ⎮  ───────
  ───────────────────────────────────────       ⎪  ───────          ⎠  6·09*
 (Simpson's Mult.)   (cub ft. to ton.)           ⎭  100872
        3        ×        35                         16812
                                                     117684
                                                     134496
                                                     ───────
 *6·09 × 28·6 (Com. Int.)                         1465·33392 tons m'l'd dis p't.
   =174·2 Centre of Buoy
 forward of No. 1 Ordinate.
```

The integrator, another and still more ingenious instrument, by M. J. Amsler-Laffon, was invented theoretically shortly after the plainimeter just described (in the year 1855), but was first constructed for practical use in the year 1867, the first instrument made being exhibited in the Paris International Exhibition in the year named. It was not introduced into England till the year 1878, and although adapted for other uses than those

involved in scientific calculations connected with shipbuilding·
it was in this connection that attention was first seriously
directed towards it. In 1880 the late Mr. C. W. Merrifield·
described the instrument, and traced the mathematical principles upon which it is based, before the Institution of Naval
Architects, and in 1882, before the same body, Mr. J. H. Biles,
naval architect for the firm of Messrs J. & G. Thomson, called
attention to the usefulness of the instrument in stability
investigations, showing by specimen calculations and other
particulars its great adaptability to this class of work, even
in the hands of youthful and untrained operators. A still
more recent and exhaustive paper devoted to the claims of the

FIG. 27.

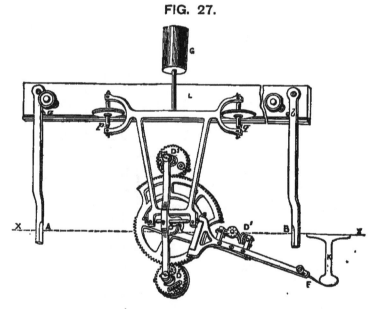

AMSLER'S MECHANICAL INTEGRATOR.

integrator upon naval architects was read before the same
Institution by Dr. A. Amsler, the son of the inventor, at its
last meeting. This paper was chiefly concerned with demon-
strating the advantages of the integrator in respect of time
saved, as well as in respect of its great accuracy.

The object of the integrator is to find at one operation the
area, the statical moment, and the moment of inertia of any
closed curve or figure by simply tracing out the curve with a

pointer, the results being réad off directly from the instrument, as in the case of the planimeter, and with a correspondingly small amount of after calculation. As shown by Fig. 25, the essential parts of the integrator are a rail L, having groove with which to guide the wheels p and q of a carriage provided with rollers D₁ D2 D3 moving on the surface of the drawing. The contour of the figure to be dealt with is traced—in the direction of the movement of the hands of a watch—by the pointer F, this pointer being attached to an arm moving on the vertical centre of the instrument while the whole mechanism runs to and fro on the rail L. Under these conditions the rollers D₁ D2 D3 execute movements partly rolling, partly sliding, and by readings taken from the divisions engraved upon their circumferences at the beginning and the end of the whole movement, together with simple arithmetical processes, the nature of which may be inferred from the explanations given of the planimeter readings, the three quantities sought are arrived at.

In a valuable appendix to the paper read by Dr. Amsler, before the Institution of Naval Architects, specimen sheets are given of several calculations, of a vessel of about 4000 tons, the forms in which the figures are entered being so arranged as to avoid all unnecessary trouble in measuring and calculating, and to contain at the same time a check on the results. The accuracy and the speed of working depend, of course, to a considerable extent on the person using the integrator, but as showing what can be obtained with the instrument after some practice, the specimens given in the paper referred to are certainly remarkable. For the calculations of the data necessary for the construction of the curves of displacement and vertical position of centre of buoyancy, the complete integrator and arithmetical work took only two hours; for the data requisite for the curve of displacement per inch immersion, and transverse meta-centre one hour was taken; and for the complete calculation, affording data to construct a stability curve, the time taken was only eight hours. A similar calculation done in the ordinary arithmetical method, and giving results far less reliable, would have taken as many days. All the work, it should be added, was

done without the aid of an assistant. Amongst other calculations besides displacement and stability in connection with
which the integrator is greatly advantageous, are those concerned with the strength of vessels and with the longitudinal
strains to which they are subject at sea through unequal
distributions of weight and buoyancy, already fully referred
to in the chapter on scientific progress.

BENNETT & THOMSON, PRINTERS.

BOOKS RELATING

TO

APPLIED SCIENCE

PUBLISHED BY

E. & F. N. SPON,

LONDON: 125, STRAND.

NEW YORK: 35, MURRAY STREET.

———•———

A Pocket-Book for Chemists, Chemical Manufacturers,
Metallurgists, Dyers, Distillers, Brewers, Sugar Refiners, Photographers,
Students, etc., etc. By THOMAS BAYLEY, Assoc. R.C. Sc. Ireland, Ana.
lytical and Consulting Chemist and Assayer. Third edition, with
additions, 437 pp., royal 32mo, roan, gilt edges, 5*s*.

SYNOPSIS OF CONTENTS :

Atomic Weights and Factors—Useful Data—Chemical Calculations—Rules for Indirect
Analysis—Weights and Measures—Thermometers and Barometers—Chemical Physics—
Boiling Points, etc.—Solubility of Substances—Methods of Obtaining Specific Gravity—Con-
version of Hydrometers—Strength of Solutions by Specific Gravity—Analysis—Gas Analysis—
Water Analysis—Qualitative Analysis and Reactions—Volumetric Analysis—Manipulation—
Mineralogy — Assaying — Alcohol — Beer — Sugar — Miscellaneous Technological matter
relating to Potash, Soda, Sulphuric Acid, Chlorine, Tar Products, Petroleum, Milk, Tallow,
Photography, Prices, Wages, Appendix, etc., etc.

The Mechanician: A Treatise on the Construction
and Manipulation of Tools, for the use and instruction of Young Engineers
and Scientific Amateurs, comprising the Arts of Blacksmithing and Forg.
ing ; the Construction and Manufacture of Hand Tools, and the various
Methods of Using and Grinding them ; the Construction of Machine Tools,
and how to work them ; Machine Fitting and Erection ; description of
Hand and Machine Processes ; Turning and Screw Cutting ; principles of
Constructing and details of Making and Erecting Steam Engines, and the
various details of setting out work, etc., etc. By CAMERON KNIGHT,
Engineer. *Containing* 1147 *illustrations*, and 397 pages of letter-press.
Third edition, 4to, cloth, 18*s*.

B

On Designing Belt Gearing. By E. J. COWLING WELCH, Mem. Inst. Mech. Engineers, Author of 'Designing Valve Gearing.' Fcap. 8vo, sewed, 6*d.*

A Handbook of Formulæ, Tables, and Memoranda, for Architectural Surveyors and others engaged in Building. By J. T. HURST, C.E. Thirteenth edition, royal 32mo, roan, 5*s.*

"It is no disparagement to the many excellent publications we refer to, to say that in our opinion this little pocket-book of Hurst's is the very best of them all, without any exception. It would be useless to attempt a recapitulation of the contents, for it appears to contain almost *everything* that anyone connected with building could require, and, best of all, made up in a compact form for carrying in the pocket, measuring only 5 in. by 3 in., and about ¼ in. thick, in a limp cover. We congratulate the author on the success of his laborious and practically compiled little book, which has received unqualified and deserved praise from every professional person to whom we have shown it."—*The Dublin Builder.*

Tabulated Weights of Angle, Tee, Bulb, Round, Square, and Flat Iron and Steel, and other information for the use of Naval Architects and Shipbuilders. By C. H. JORDAN, M.I.N.A. Fourth edition, 32mo, cloth, 2*s.* 6*d.*

Quantity Surveying. By J. LEANING. With 42 illustrations, crown 8vo, cloth, 9*s.*

CONTENTS :

A complete Explanation of the London Practice.
General Instructions.
Order of Taking Off.
Modes of Measurement of the various Trades.
Use and Waste.
Ventilation and Warming.
Credits, with various Examples of Treatment.
Abbreviations.
Squaring the Dimensions.
Abstracting, with Examples in illustration of each Trade.
Billing.
Examples of Preambles to each Trade.
Form for a Bill of Quantities.
Do. Bill of Credits.
Do. Bill for Alternative Estimate.
Restorations and Repairs, and Form of Bill.
Variations before Acceptance of Tender.
Errors in a Builder's Estimate.

Schedule of Prices.
Form of Schedule of Prices.
Analysis of Schedule of Prices.
Adjustment of Accounts.
Form of a Bill of Variations.
Remarks on Specifications.
Prices and Valuation of Work, with Examples and Remarks upon each Trade.
The Law as it affects Quantity Surveyors, with Law Reports.
Taking Off after the Old Method.
Northern Practice.
The General Statement of the Methods recommended by the Manchester Society of Architects for taking Quantities.
Examples of Collections.
Examples of "Taking Off" in each Trade.
Remarks on the Past and Present Methods of Estimating.

A Practical Treatise on Heat, as applied to the Useful Arts; for the Use of Engineers, Architects, &c. By THOMAS BOX. With 14 plates. Third edition, crown 8vo, cloth, 12*s.* 6*d.*

A Descriptive Treatise on Mathematical Drawing Instruments: their construction, uses, qualities, selection, preservation, and suggestions for improvements, with hints upon Drawing and Colouring. By W. F. STANLEY, M.R.I. Fifth edition, *with numerous illustrations,* crown 8vo, cloth, 5*s.*

Spons' Architects' and Builders' Pocket-Book of Prices and Memoranda. Edited by W. YOUNG, Architect. Royal 32mo, roan, 4s. 6d. ; or cloth, red edges, 3s. 6d. *Published annually.* Eleventh edition. *Now ready.*

Long-Span Railway Bridges, comprising Investigations of the Comparative Theoretical and Practical Advantages of the various adopted or proposed Type Systems of Construction, with numerous Formulæ and Tables giving the weight of Iron or Steel required in Bridges from 300 feet to the limiting Spans ; to which are added similar Investigations and Tables relating to Short-span Railway Bridges. Second and revised edition. By B. BAKER, Assoc. Inst. C.E. *Plates*, crown 8vo, cloth, 5s.

Elementary Theory and Calculation of Iron Bridges and Roofs. By AUGUST RITTER, Ph.D., Professor at the Polytechnic School at Aix-la-Chapelle. Translated from the third German edition, by H. R. SANKEY, Capt. R.E. With 500 *illustrations*, 8vo, cloth, 15s.

The Builder's Clerk : a Guide to the Management of a Builder's Business. By THOMAS BALES. Fcap. 8vo, cloth, 1s. 6d.

The Elementary Principles of Carpentry. By THOMAS TREDGOLD. Revised from the original edition, and partly re-written, by JOHN THOMAS HURST. Contained in 517 pages of letter-press, and *illustrated with 48 plates and 150 wood engravings.* Third edition, crown 8vo, cloth, 18s.

Section I. On the Equality and Distribution of Forces — Section II. Resistance of Timber — Section III. Construction of Floors — Section IV. Construction of Roofs — Section V. Construction of Domes and Cupolas — Section VI. Construction of Partitions — Section VII. Scaffolds, Staging, and Gantries — Section VIII. Construction of Centres for Bridges — Section IX. Coffer-dams, Shoring, and Strutting — Section X. Wooden Bridges and Viaducts — Section XI. Joints, Straps, and other Fastenings — Section XII. Timber.

Our Factories, Workshops, and Warehouses : their Sanitary and Fire-Resisting Arrangements. By B. H. THWAITE, Assoc. Mem. Inst. C.E. *With* 183 *wood engravings,* crown 8vo, cloth, 9s.

Gold : Its Occurrence and Extraction, embracing the Geographical and Geological Distribution and the Mineralogical Characters of Gold-bearing rocks ; the peculiar features and modes of working Shallow Placers, Rivers, and Deep Leads ; Hydraulicing ; the Reduction and Separation of Auriferous Quartz ; the treatment of complex Auriferous ores containing other metals ; a Bibliography of the subject and a Glossary of Technical and Foreign Terms. By ALFRED G. LOCK, F.R.G.S. *With numerous illustrations and maps,* 1250 pp., super-royal 8vo, cloth, 2l. 12s. 6d.

B 2

A Practical Treatise on Coal Mining. By GEORGE
G. ANDRÉ, F.G.S., Assoc. Inst. C.E., Member of the Society of Engineers.
With 82 *lithographic plates.* 2 vols., royal 4to, cloth, 3*l.* 12*s.*

Iron Roofs: Examples of Design, Description. *Illustrated with* 64 *Working Drawings of Executed Roofs.* By ARTHUR T.
WALMISLEY, Assoc. Mem. Inst. C.E. Imp. 4to, half-morocco, £2 12*s.* 6*d.*

A History of Electric Telegraphy, to the Year 1837.
Chiefly compiled from Original Sources, and hitherto Unpublished Docu-
ments, by J. J. FAHIE, Mem. Soc. of Tel. Engineers, and of the International Society of Electricians, Paris. Crown 8vo, cloth, 9*s.*

Spons' Information for Colonial Engineers. Edited
by J. T. HURST. Demy 8vo, sewed.

No. 1, Ceylon. By ABRAHAM DEANE, C.E. 2*s.* 6*d.*

CONTENTS:

Introductory Remarks—Natural Productions—Architecture and Engineering—Topo-
graphy, Trade, and Natural History—Principal Stations—Weights and Measures, etc., etc.

No. 2. Southern Africa, including the Cape Colony, Natal, and the
Dutch Republics. By HENRY HALL, F.R.G.S., F.R.C.I. With
Map. 3*s.* 6*d.*

CONTENTS:

General Description of South Africa—Physical Geography with reference to Engineering
Operations—Notes on Labour and Material in Cape Colony—Geological Notes on Rock
Formation in South Africa—Engineering Instruments for Use in South Africa—Principal
Public Works in Cape Colony: Railways, Mountain Roads and Passes, Harbour Works,
Bridges, Gas Works, Irrigation and Water Supply, Lighthouses, Drainage and Sanitary
Engineering, Public Buildings, Mines—Table of Woods in South Africa—Animals used for
Draught Purposes—Statistical Notes—Table of Distances—Rates of Carriage, etc.

No. 3. India. By F. C. DANVERS, Assoc. Inst. C.E. With Map. 4*s.* 6*d.*

CONTENTS:

Physical Geography of India—Building Materials—Roads—Railways—Bridges—Irrigation—River Works—Harbours—Lighthouse Buildings—Native Labour—The Principal
Trees of India—Money—Weights and Measures—Glossary of Indian Terms, etc.

A Practical Treatise on Casting and Founding,
including descriptions of the modern machinery employed in the art. By
N. E. SPRETSON, Engineer. Third edition, with 82 *plates* drawn to
scale, 412 pp., demy 8vo, cloth, 18*s.*

Steam Heating for Buildings; or, Hints to Steam
Fitters, being a description of Steam Heating Apparatus for Warming
and Ventilating Private Houses and Large Buildings, with remarks on
Steam, Water, and Air in their relation to Heating. By W. J. BALDWIN.
With many illustrations. Fourth edition, crown 8vo, cloth 10*s.* 6*d.*

The Depreciation of Factories and their Valuation.
By EWING MATHESON, M. Inst. C.E. 8vo, cloth, 6s.

A Handbook of Electrical Testing. By H. R. KEMPE,
M.S.T.E. Third edition, revised and enlarged, crown 8vo, cloth, 15s.

Gas Works: their Arrangement, Construction, Plant,
and Machinery. By F. COLYER, M. Inst. C.E. *With* 31 *folding plates,*
8vo, cloth, 24s.

The Clerk of Works: a Vade-Mecum for all engaged
in the Superintendence of Building Operations. By G. G. HOSKINS,
F.R.I.B.A. Third edition, fcap. 8vo, cloth, 1s. 6d.

American Foundry Practice: Treating of Loam,
Dry Sand, and Green Sand Moulding, and containing a Practical Treatise
upon the Management of Cupolas, and the Melting of Iron. By T. D.
WEST, Practical Iron Moulder and Foundry Foreman. Second edition,
with numerous illustrations, crown 8vo, cloth, 10s. 6d.

The Maintenance of Macadamised Roads. By T.
CODRINGTON, M.I.C.E, F.G.S., General Superintendent of County Roads
for South Wales. 8vo, cloth, 6s.

*Hydraulic Steam and Hand Power Lifting and
Pressing Machinery.* By FREDERICK COLYER, M. Inst. C.E., M. Inst. M.E.
With 73 *plates,* 8vo, cloth, 18s.

Pumps and Pumping Machinery. By F. COLYER,
M.I.C.E., M.I.M.E. *With* 23 *folding plates,* 8vo, cloth, 12s. 6d.

The Municipal and Sanitary Engineer's Handbook.
By H. PERCY BOULNOIS, Mem. Inst. C.E., Borough Engineer, Portsmouth. *With numerous illustrations,* demy 8vo, cloth, 12s. 6d.

CONTENTS:

The Appointment and Duties of the Town Surveyor—Traffic—Macadamised Roadways—
Steam Rolling—Road Metal and Breaking—Pitched Pavements—Asphalte—Wood Pavements
—Footpaths—Kerbs and Gutters—Street Naming and Numbering—Street Lighting—Sewer-
age—Ventilation of Sewers—Disposal of Sewage—House Drainage—Disinfection—Gas and
Water Companies, &c., Breaking up Streets—Improvement of Private Streets—Borrowing
Powers—Artizans' and Labourers' Dwellings—Public Conveniences—Scavenging, including
Street Cleansing—Watering and the Removing of Snow—Planting Street Trees—Deposit of
Plans—Dangerous Buildings—Hoardings—Obstructions—Improving Street Lines—Cellar
Openings—Public Pleasure Grounds—Cemeteries—Mortuaries—Cattle and Ordinary Markets
—Public Slaughter-houses, etc.—Giving numerous Forms of Notices, Specifications, and
General Information upon these and other subjects of great importance to Municipal Engi-
neers and others engaged in Sanitary Work.

*Tables of the Principal Speeds occurring in Mechanical
Engineering*, expressed in metres in a second. By P. KEERAYEFF, Chief
Mechanic of the Obouchoff Steel Works, St. Petersburg; translated by
SERGIUS KERN, M.E. Fcap. 8vo, sewed, 6*d*.

*A Treatise on the Origin, Progress, Prevention, and
Cure of Dry Rot in Timber;* with Remarks on the Means of Preserving
Wood from Destruction by Sea-Worms, Beetles, Ants, etc. By THOMAS
ALLEN BRITTON, late Surveyor to the Metropolitan Board of Works,
etc., etc. *With* 10 *plates*, crown 8vo, cloth, 7*s*. 6*d*.

Metrical Tables. By G. L. MOLESWORTH, M.I.C.E.
32mo, cloth, 1*s*. 6*d*.

CONTENTS.

General—Linear Measures—Square Measures—Cubic Measures—Measures of Capacity—
Weights—Combinations—Thermometers.

Elements of Construction for Electro-Magnets. By
Count TH. DU MONCEL, Mem. de l'Institut de France. Translated from
the French by C. J. WHARTON. Crown 8vo, cloth, 4*s*. 6*d*.

Electro-Telegraphy. By FREDERICK S. BEECHEY,
Telegraph Engineer. A Book for Beginners. *Illustrated.* Fcap. 8vo,
sewed, 6*d*.

Handrailing: by the Square Cut. By JOHN JONES,
Staircase Builder. Fourth edition, *with seven plates*, 8vo, cloth, 3*s*. 6*d*.

Handrailing: by the Square Cut. By JOHN JONES,
Staircase Builder. Part Second, *with eight plates*, 8vo, cloth, 3*s*. 6*d*.

Practical Electrical Units Popularly Explained, with
numerous illustrations and Remarks. By JAMES SWINBURNE, late of
J. W. Swan and Co., Paris, late of Brush-Swan Electric Light Company,
U.S.A. 18mo, cloth, 1*s*. 6*d*.

Philipp Reis, Inventor of the Telephone: A Biographical
Sketch. With Documentary Testimony, Translations of the Original
Papers of the Inventor, &c. By SILVANUS P. THOMPSON, B.A., Dr. Sc.,
Professor of Experimental Physics in University College, Bristol. *With
illustrations,* 8vo, cloth, 7*s*. 6*d*.

*A Treatise on the Use of Belting for the Transmis-
sion of Power.* By J. H. COOPER. Second edition, *illustrated*, 8vo,
cloth, 15*s*.

A Pocket-Book of Useful Formulæ and Memoranda

for Civil and Mechanical Engineers. By GUILFORD L. MOLESWORTH, Mem. Inst. C.E., Consulting Engineer to the Government of India for State Railways. With numerous illustrations, 744 pp. Twenty-first edition, revised and enlarged, 32mo, roan, 6s.

SYNOPSIS OF CONTENTS:

Surveying, Levelling, etc.—Strength and Weight of Materials—Earthwork, Brickwork, Masonry, Arches, etc.—Struts, Columns, Beams, and Trusses—Flooring, Roofing, and Roof Trusses—Girders, Bridges, etc.—Railways and Roads—Hydraulic Formulæ—Canals, Sewers, Waterworks, Docks—Irrigation and Breakwaters—Gas, Ventilation, and Warming—Heat, Light, Colour, and Sound—Gravity: Centres, Forces, and Powers—Millwork, Teeth of Wheels, Shafting, etc.—Workshop Recipes—Sundry Machinery—Animal Power—Steam and the Steam Engine—Water-power, Water-wheels, Turbines, etc.—Wind and Windmills—Steam Navigation, Ship Building, Tonnage, etc.—Gunnery, Projectiles, etc.—Weights, Measures, and Money—Trigonometry, Conic Sections, and Curves—Telegraphy—Mensuration—Tables of Areas and Circumference, and Arcs of Circles—Logarithms, Square and Cube Roots, Powers—Reciprocals, etc.—Useful Numbers—Differential and Integral Calculus—Algebraic Signs—Telegraphic Construction and Formulæ.

Spons' Tables and Memoranda for Engineers;

selected and arranged by J. T. HURST, C.E., Author of 'Architectural Surveyors' Handbook,' 'Hurst's Tredgold's Carpentry,' etc. Fifth edition, 64mo, roan, gilt edges, 1s. ; or in cloth case, 1s. 6d.

This work is printed in a pearl type, and is so small, measuring only 2½ in. by 1¾ in. by ⅜ in. thick, that it may be easily carried in the waistcoat pocket.

"It is certainly an extremely rare thing for a reviewer to be called upon to notice a volume measuring but 2½ in. by 1¾ in., yet these dimensions faithfully represent the size of the handy little book before us. The Volume—which contains 118 printed pages, besides a few blank pages for memoranda—is, in fact, a true pocket-book, adapted for being carried in the waistcoat pocket, and containing a far greater amount and variety of information than most people would imagine could be compressed into so small a space. . . . . The little volume has been compiled with considerable care and judgment, and we can cordially recommend it to our readers as a useful little pocket companion."—Engineering.

A Practical Treatise on Natural and Artificial

Concrete, its Varieties and Constructive Adaptations. By HENRY REID, Author of the 'Science and Art of the Manufacture of Portland Cement.' New Edition, with 59 woodcuts and 5 plates, 8vo, cloth, 15s.

Hydrodynamics: Treatise relative to the Testing of

Water-Wheels and Machinery, with various other matters pertaining to Hydrodynamics. By JAMES EMERSON. With numerous illustrations, 360 pp. Third edition, crown 8vo, cloth, 4s. 6d.

Electricity as a Motive Power. By Count TH. DU

MONCEL, Membre de l'Institut de France, and FRANK GERALDY, Ingénieur des Ponts et Chaussées. Translated and Edited, with Additions, by C. J. WHARTON, Assoc. Soc. Tel. Eng. and Elec. With 113 engravings and diagrams, crown 8vo, cloth, 7s. 6d.

Hints on Architectural Draughtsmanship. By G. W.

TUXFORD HALLATT. Fcap. 8vo, cloth, 1s. 6d.

Treatise on Valve-Gears, with special consideration of the Link-Motions of Locomotive Engines. By Dr. GUSTAV ZEUNER, Professor of Applied Mechanics at the Confederated Polytechnikum of Zurich. Translated from the Fourth German Edition, by Professor J. F. KLEIN, Lehigh University, Bethlehem, Pa. *Illustrated*, 8vo, cloth, 12s. 6d.

The French-Polisher's Manual. By a French-Polisher; containing Timber Staining, Washing, Matching, Improving, Painting, Imitations, Directions for Staining, Sizing, Embodying, Smoothing, Spirit Varnishing, French-Polishing, Directions for Re-polishing. Third edition, royal 32mo, sewed, 6d.

Hops, their Cultivation, Commerce, and Uses in various Countries. By P. L. SIMMONDS. Crown 8vo, cloth, 4s. 6d.

A Practical Treatise on the Manufacture and Distribution of Coal Gas. By WILLIAM RICHARDS. Demy 4to, with *numerous wood engravings and 29 plates*, cloth, 28s.

SYNOPSIS OF CONTENTS :

Introduction — History of Gas Lighting — Chemistry of Gas Manufacture, by Lewis Thompson, Esq., M.R.C.S. — Coal, with Analyses, by J. Paterson, Lewis Thompson, and G. R. Hislop, Esqrs. — Retorts, Iron and Clay — Retort Setting — Hydraulic Main — Condensers — Exhausters — Washers and Scrubbers — Purifiers — Purification — History of Gas Holder — Tanks, Brick and Stone, Composite, Concrete, Cast-iron, Compound Annular Wrought-iron — Specifications — Gas Holders — Station Meter — Governor — Distribution — Mains — Gas Mathematics, or Formulæ for the Distribution of Gas, by Lewis Thompson, Esq. — Services — Consumers' Meters — Regulators — Burners — Fittings — Photometer — Carburization of Gas — Air Gas and Water Gas — Composition of Coal Gas, by Lewis Thompson, Esq. — Analyses of Gas — Influence of Atmospheric Pressure and Temperature on Gas — Residual Products — Appendix — Description of Retort Settings, Buildings, etc., etc.

Practical Geometry, Perspective, and Engineering Drawing; a Course of Descriptive Geometry adapted to the Requirements of the Engineering Draughtsman, including the determination of cast shadows and Isometric Projection, each chapter being followed by numerous examples ; to which are added rules for Shading, Shade-lining, etc., together with practical instructions as to the Lining, Colouring, Printing, and general treatment of Engineering Drawings, with a chapter on drawing Instruments. By GEORGE S. CLARKE, Capt. R.E. Second edition, *with 21 plates.* 2 vols., cloth, 10s. 6d.

The Elements of Graphic Statics. By Professor KARL VON OTT, translated from the German by G. S. CLARKE, Capt. R.E., Instructor in Mechanical Drawing, Royal Indian Engineering College. *With 93 illustrations*, crown 8vo, cloth, 5s.

The Principles of Graphic Statics. By GEORGE SYDENHAM CLARKE, Capt. Royal Engineers. *With 112 illustrations.* 4to, cloth, 12s. 6d.

Dynamo-Electric Machinery : A Manual for Students of Electro-technics. By SILVANUS P. THOMPSON, B.A., D.Sc., Professor of Experimental Physics in University College, Bristol, etc., etc. *Illustrated*, 8vo, cloth, 12s. 6d.

The New Formula for Mean Velocity of Discharge
of Rivers and Canals. By W. R. KUTTER. Translated from articles in
the 'Cultur-Ingénieur,' by LOWIS D'A. JACKSON, Assoc. Inst. C.E.
8vo, cloth, 12s. 6d.

Practical Hydraulics; a Series of Rules and Tables
for the use of Engineers, etc., etc. By THOMAS BOX. Fifth edition,
numerous plates, post 8vo, cloth, 5s.

A Practical Treatise on the Construction of Hori-
zontal and Vertical Waterwheels, specially designed for the use of operative mechanics. By WILLIAM CULLEN, Millwright and Engineer. *With*
11 *plates.* Second edition, revised and enlarged, small 4to, cloth, 12s. 6d.

Tin: Describing the Chief Methods of Mining,
Dressing and Smelting it abroad ; with Notes upon Arsenic, Bismuth and
Wolfram. By ARTHUR G. CHARLETON, Mem. American Inst. of
Mining Engineers. *With plates*, 8vo, cloth, 12s. 6d.

Perspective, Explained and Illustrated. By G. S.
CLARKE, Capt. R.E. *With illustrations*, 8vo, cloth, 3s. 6d.

The Essential Elements of Practical Mechanics;
based on the Principle of Work, designed for Engineering Students. By
OLIVER BYRNE, formerly Professor of Mathematics, College for Civil
Engineers. Third edition, *with* 148 *wood engravings*, post 8vo, cloth,
7s. 6d.

CONTENTS:

Chap. 1. How Work is Measured by a Unit, both with and without reference to a Unit
of Time—Chap. 2. The Work of Living Agents, the Influence of Friction, and introduces
one of the most beautiful Laws of Motion—Chap. 3. The principles expounded in the first and
second chapters are applied to the Motion of Bodies—Chap. 4. The Transmission of Work by
simple Machines—Chap. 5. Useful Propositions and Rules.

The Practical Millwright and Engineer's Ready
Reckoner; or Tables for finding the diameter and power of cog-wheels,
diameter, weight, and power of shafts, diameter and strength of bolts, etc.
By THOMAS DIXON. Fourth edition, 12mo, cloth, 3s.

Breweries and Maltings: their Arrangement, Con-
struction, Machinery, and Plant. By G. SCAMELL, F.R.I.B.A. Second
edition, revised, enlarged, and partly rewritten. By F. COLYER, M.I.C.E.,
M.I.M.E. *With* 20 *plates*, 8vo, cloth, 18s.

A Practical Treatise on the Manufacture of Starch,
Glucose, Starch-Sugar, and Dextrine, based on the German of L. Von
Wagner, Professor in the Royal Technical School, Buda Pesth, and
other authorities. By JULIUS FRANKEL ; edited by ROBERT HUTTER,
proprietor of the Philadelphia Starch Works. *With* 58 *illustrations*,
344 pp., 8vo, cloth, 18s.

A Practical Treatise on Mill-gearing, Wheels, Shafts,
Riggers, etc.; for the use of Engineers. By THOMAS BOX. Third
edition, *with 11 plates.* Crown 8vo, cloth, 7s. 6d.

Mining Machinery: a Descriptive Treatise on the
Machinery, Tools, and other Appliances used in Mining. By G. G.
ANDRÉ, F.G.S., Assoc. Inst. C.E., Mem. of the Society of Engineers.
Royal 4to, uniform with the Author's Treatise on Coal Mining, con-
taining 182 *plates,* accurately drawn to scale, with descriptive text, in
2 vols., cloth, 3l. 12s.

CONTENTS :

Machinery for Prospecting, Excavating, Hauling, and Hoisting—Ventilation—Pumping—
Treatment of Mineral Products, including Gold and Silver, Copper, Tin, and Lead, Iron,
Coal, Sulphur, China Clay, Brick Earth, etc.

Tables for Setting out Curves for Railways, Canals,
Roads, etc., varying from a radius of five chains to three miles. By A.
KENNEDY and R. W. HACKWOOD. *Illustrated,* 32mo, cloth, 2s. 6d.

The Science and Art of the Manufacture of Portland
Cement, with observations on some of its constructive applications. *With*
66 *illustrations.* By HENRY REID, C.E., Author of 'A Practical
Treatise on Concrete,' etc., etc. 8vo, cloth, 18s.

The Draughtsman's Handbook of Plan and Map
Drawing; including instructions for the preparation of Engineering,
Architectural, and Mechanical Drawings. *With numerous illustrations*
in the text, and 33 plates (15 printed in colours). By G. G. ANDRÉ,
F.G.S., Assoc. Inst. C.E. 4to, cloth, 9s.

CONTENTS :

The Drawing Office and its Furnishings—Geometrical Problems—Lines, Dots, and their
Combinations—Colours, Shading, Lettering, Bordering, and North Points—Scales—Plotting
—Civil Engineers' and Surveyors' Plans—Map Drawing—Mechanical and Architectural
Drawing—Copying and Reducing Trigonometrical Formulæ, etc., etc.

The Boiler-maker's and Iron Ship-builder's Companion,
comprising a series of original and carefully calculated tables, of the
utmost utility to persons interested in the iron trades. By JAMES FODEN,
author of 'Mechanical Tables,' etc. Second edition revised, *with illustra-*
tions, crown 8vo, cloth, 5s.

Rock Blasting: a Practical Treatise on the means
employed in Blasting Rocks for Industrial Purposes. By G. G. ANDRÉ,
F.G.S., Assoc. Inst. C.E. *With 56 illustrations and 12 plates,* 8vo, cloth,
10s. 6d.

Painting and Painters' Manual: a Book of Facts
for Painters and those who Use or Deal in Paint Materials. By C. L.
CONDIT and J. SCHELLER. *Illustrated,* 8vo, cloth, 10s. 6d.

A Treatise on Ropemaking as practised in public and private Rope-yards, with a Description of the Manufacture, Rules, Tables of Weights, etc., adapted to the Trade, Shipping, Mining, Railways, Builders, etc. By R. CHAPMAN, formerly foreman to Messrs. Huddart and Co., Limehouse, and late Master Ropemaker to H.M. Dockyard, Deptford. Second edition, 12mo, cloth, 3s.

Laxton's Builders' and Contractors' Tables ; for the use of Engineers, Architects, Surveyors, Builders, Land Agents, and others. Bricklayer, containing 22 tables, with nearly 30,000 calculations. 4to, cloth, 5s.

Laxton's Builders' and Contractors' Tables. Excavator, Earth, Land, Water, and Gas, containing 53 tables, with nearly 24,000 calculations. 4to, cloth, 5s.

Sanitary Engineering: a Guide to the Construction of Works of Sewerage and House Drainage, with Tables for facilitating the calculations of the Engineer. By BALDWIN LATHAM, C.E., M. Inst. C.E., F.G.S., F.M.S., Past-President of the Society of Engineers. Second edition, *with numerous plates and woodcuts*, 8vo, cloth, 1l. 10s.

Screw Cutting Tables for Engineers and Machinists, giving the values of the different trains of Wheels required to produce Screws of any pitch, calculated by Lord Lindsay, M.P., F.R.S., F.R.A.S., etc. Cloth, oblong, 2s.

Screw Cutting Tables, for the use of Mechanical Engineers, showing the proper arrangement of Wheels for cutting the Threads of Screws of any required pitch, with a Table for making the Universal Gas-pipe Threads and Taps. By W. A. MARTIN, Engineer. Second edition, oblong, cloth, 1s., or sewed, 6d.

A Treatise on a Practical Method of Designing Slide-Valve Gears by Simple Geometrical Construction, based upon the principles enunciated in Euclid's Elements, and comprising the various forms of Plain Slide-Valve and Expansion Gearing ; together with Stephenson's, Gooch's, and Allan's Link-Motions, as applied either to reversing or to variable expansion combinations. By EDWARD J. COWLING WELCH, Memb. Inst. Mechanical Engineers. Crown 8vo, cloth, 6s.

Cleaning and Scouring : a Manual for Dyers, Laundresses, and for Domestic Use. By S. CHRISTOPHER. 18mo, sewed, 6d.

A Handbook of House Sanitation ; for the use of all persons seeking a Healthy Home. A reprint of those portions of Mr. Bailey-Denton's Lectures on Sanitary Engineering, given before the School of Military Engineering, which related to the "Dwelling," enlarged and revised by his Son, E. F. BAILEY-DENTON, C.E., B.A. *With* 140 *illustrations*, 8vo, cloth, 8s. 6d.

A Glossary of Terms used in Coal Mining. By
WILLIAM STUKELEY GRESLEY, Assoc. Mem. Inst. C.E., F.G.S., Member
of the North of England Institute of Mining Engineers. *Illustrated with
numerous woodcuts and diagrams*, crown 8vo, cloth, 5s.

A Pocket-Book for Boiler Makers and Steam Users,
comprising a variety of useful information for Employer and Workman,
Government Inspectors, Board of Trade Surveyors, Engineers in charge
of Works and Slips, Foremen of Manufactories, and the general Steam-
using Public. By MAURICE JOHN SEXTON. Second edition, royal
32mo, roan, gilt edges, 5s.

The Strains upon Bridge Girders and Roof Trusses,
including the Warren, Lattice, Trellis, Bowstring, and other Forms of
Girders, the Curved Roof, and Simple and Compound Trusses. By
THOS. CARGILL, C.E.B.A.T., C.D., Assoc. Inst. C.E., Member of the
Society of Engineers. *With 64 illustrations, drawn and worked out to scale*,
8vo, cloth, 12s. 6d.

A Practical Treatise on the Steam Engine, containing Plans and Arrangements of Details for Fixed Steam Engines,
with Essays on the Principles involved in Design and Construction. By
ARTHUR RIGG, Engineer, Member of the Society of Engineers and of
the Royal Institution of Great Britain. Demy 4to, *copiously illustrated
with woodcuts and 96 plates*, in one Volume, half-bound morocco, 2l. 2s.;
or cheaper edition, cloth, 25s.

This work is not, in any sense, an elementary treatise, or history of the steam engine, but
is intended to describe examples of Fixed Steam Engines without entering into the wide
domain of locomotive or marine practice. To this end illustrations will be given of the most
recent arrangements of Horizontal, Vertical, Beam, Pumping, Winding, Portable, Semi-
portable, Corliss, Allen, Compound, and other similar Engines, by the most eminent Firms in
Great Britain and America. The laws relating to the action and precautions to be observed
in the construction of the various details, such as Cylinders, Pistons, Piston-rods, Connecting-
rods, Cross-heads, Motion-blocks, Eccentrics, Simple, Expansion, Balanced, and Equilibrium
Slide-Valves, and Valve-gearing will be minutely dealt with. In this connection will be found
articles upon the Velocity of Reciprocating Parts and the Mode of Applying the Indicator,
Heat and Expansion of Steam Governors, and the like. It is the writer's desire to draw
illustrations from every possible source, and give only those rules that present practice deems
correct.」

Barlow's Tables of Squares, Cubes, Square Roots,
Cube Roots, Reciprocals of all Integer Numbers up to 10,000. Post 8vo,
cloth, 6s.

Camus (M.) Treatise on the Teeth of Wheels, demon-
strating the best forms which can be given to them for the purposes of
Machinery, such as Mill-work and Clock-work, and the art of finding
their numbers. Translated from the French, with details of the present
practice of Millwrights, Engine Makers, and other Machinists, by
ISAAC HAWKINS. Third edition, *with 18 plates*, 8vo, cloth, 5s.

A Practical Treatise on the Science of Land and Engineering Surveying, Levelling, Estimating Quantities, etc., with a general description of the several Instruments required for' Surveying, Levelling, Plotting, etc. By H. S. MERRETT. Third edition, 41 *plates with illustrations and tables*, royal 8vo, cloth, 12s. 6d.

PRINCIPAL CONTENTS :

Part 1. Introduction and the Principles of Geometry. Part 2. Land Surveying: comprising General Observations—The Chain—Offsets Surveying by the Chain only—Surveying Hilly Ground—To Survey an Estate or Parish by the Chain only—Surveying with the Theodolite—Mining and Town Surveying—Railroad Surveying—Mapping—Division and Laying out of Land—Observations on Enclosures—Plane Trigonometry. Part 3. Levelling—Simple and Compound Levelling—The Level Book—Parliamentary Plan and Section—Levelling with a Theodolite—Gradients—Wooden Curves—To Lay out a Railway Curve—Setting out Widths. Part 4. Calculating Quantities generally for Estimates—Cuttings and Embankments—Tunnels—Brickwork—Ironwork—Timber Measuring. Part 5. Description and Use of Instruments in Surveying and Plotting—The Improved Dumpy Level—Troughton's Level—The Prismatic Compass—Proportional Compass—Box Sextant—Vernier—Pantagraph—Merrett's Improved Quadrant—Improved Computation Scale—The Diagonal Scale—Straight Edge and Sector. Part 6. Logarithms of Numbers—Logarithmic Sines and Co-Sines, Tangents and Co-Tangents—Natural Sines and Co-Sines—Tables for Earthwork, for Setting out Curves, and for various Calculations, etc., etc., etc.

Saws : the History, Development, Action, Classification, and Comparison of Saws of all kinds. By ROBERT GRIMSHAW. *With* 220 *illustrations*, 4to, cloth, 12s. 6d.

A Supplement to the above ; containing additional practical matter, more especially relating to the forms of Saw Teeth for special material and conditions, and to the behaviour of Saws under particular conditions. *With* 120 *illustrations*, cloth, 9s.

A Guide for the Electric Testing of Telegraph Cables. By Capt. V. HOSKIŒR, Royal Danish Engineers. *With illustrations*, second edition, crown 8vo, cloth, 4s. 6d.

Laying and Repairing Electric Telegraph Cables. By Capt. V. HOSKIŒR, Royal Danish Engineers. Crown 8vo, cloth, 3s. 6d.

A Pocket-Book of Practical Rules for the Proportions of Modern Engines and Boilers for Land and Marine purposes. By N. P. BURGH. Seventh edition, royal 32mo, roan, 4s. 6d.

The Assayer's Manual : an Abridged Treatise on the Docimastic Examination of Ores and Furnace and other Artificial Products. By BRUNO KERL. Translated by W. T. BRANNT. *With* 65 *illustrations*, 8vo, cloth, 12s. 6d.

The Steam Engine considered as a Heat Engine : a Treatise on the Theory of the Steam Engine, illustrated by Diagrams, Tables, and Examples from Practice. By JAS. H. COTTERILL, M.A., F.R.S., Professor of Applied Mechanics in the Royal Naval College. 8vo, cloth, 12s. 6d.

Electricity: its Theory, Sources, and Applications.
By J. T. SPRAGUE, M.S.T.E. Second edition, revised and enlarged, *with numerous illustrations*, crown 8vo, cloth, 15*s.*

The Practice of Hand Turning in Wood, Ivory, Shell,
etc., with Instructions for Turning such Work in Metal as may be required in the Practice of Turning in Wood, Ivory, etc. ; also an Appendix on Ornamental Turning. (A book for beginners.) By FRANCIS CAMPIN.
Third edition, *with wood engravings*, crown 8vo, cloth, 6*s.*

CONTENTS :

On Lathes—Turning Tools—Turning Wood—Drilling—Screw Cutting—Miscellaneous Apparatus and Processes—Turning Particular Forms—Staining—Polishing—Spinning Metals —Materials—Ornamental Turning, etc.

Health and Comfort in House Building, or Ventila-
tion with Warm Air by Self-Acting Suction. Power, with Review of the mode of Calculating the Draught in Hot-Air Flues, and with some actual Experiments. By J. DRYSDALE, M.D., and J. W. HAYWARD, M.D. Second edition, with Supplement, *with plates*, demy 8vo, cloth, 7*s. 6d.*

Treatise on Watchwork, Past and Present. By the
Rev. H. L. NELTHROPP, M.A., F.S.A. *With 32 illustrations,* crown 8vo, cloth, 6*s. 6d.*

CONTENTS :

Definitions of Words and Terms used in Watchwork—Tools—Time—Historical Summary—On Calculations of the Numbers for Wheels and Pinions; their Proportional Sizes, Trains, etc.—Of Dial Wheels, or Motion Work—Length of Time of Going without Winding up—The Verge—The Horizontal—The Duplex—The Lever—The Chronometer—Repeating Watches—Keyless Watches—The Pendulum, or Spiral Spring—Compensation—Jewelling of Pivot Holes—Clerkenwell—Fallacies of the Trade—Incapacity of Workmen—How to Choose and Use a Watch, etc.

Notes in Mechanical Engineering. Compiled prin-
cipally for the use of the Students attending the Classes on this subject at the City of London College. By HENRY ADAMS, Mem. Inst. M.E., Mem. Inst. C.E., Mem. Soc. of Engineers. Crown 8vo, cloth, 2*s. 6d.*

Algebra Self-Taught. By W. P. HIGGS, M.A.,
D.Sc., LL.D., Assoc. Inst. C.E., Author of ' A Handbook of the Differential Calculus,' etc. Second edition, crown 8vo, cloth, 2*s. 6d.*

CONTENTS :

Symbols and the Signs of Operation—The Equation and the Unknown Quantity—Positive and Negative Quantities—Multiplication—Involution—Exponents—Negative Exponents—Roots, and the Use of Exponents as Logarithms—Logarithms—Tables of Logarithms and Proportionate Parts—Transformation of System of Logarithms—Common Uses of Common Logarithms—Compound Multiplication and the Binomial Theorem—Division, Fractions, and Ratio—Continued Proportion—The Series and the Summation of the Series—Limit of Series—Square and Cube Roots—Equations—List of Formulæ, etc.

Spons' Dictionary of Engineering, Civil, Mechanical,
Military, and Naval; with technical terms in French, German, Italian, and Spanish, 3100 pp., and *nearly* 8000 *engravings*, in super-royal 8vo, in 8 divisions, 5*l.* 8*s.* Complete in 3 vols., cloth, 5*l.* 5*s.* Bound in a superior manner, half-morocco, top edge gilt, 3 vols., 6*l.* 12*s.*

In super-royal 8vo, 1168 pp., *with* 2400 *illustrations*, in 3 Divisions, cloth, price 13*s.* 6*d.* each ; or 1 vol., cloth, 2*l.* ; or half-morocco, 2*l.* 8*s.*

A SUPPLEMENT

TO

SPONS' DICTIONARY OF ENGINEERING.

EDITED BY ERNEST SPON, MEMB. SOC. ENGINEERS.

Abacus, Counters, Speed Indicators, and Slide Rule.

Agricultural Implements and Machinery.

Air Compressors.

Animal Charcoal Machinery.

Antimony.

Axles and Axle-boxes.

Barn Machinery.

Belts and Belting.

Blasting. Boilers.

Brakes.

Brick Machinery.

Bridges.

Cages for Mines.

Calculus, Differential and Integral.

Canals.

Carpentry.

Cast Iron.

Cement, Concrete, Limes, and Mortar.

Chimney Shafts.

Coal Cleansing and Washing.

Coal Mining.

Coal Cutting Machines.

Coke Ovens. Copper.

Docks. Drainage.

Dredging Machinery.

Dynamo - Electric and Magneto-Electric Machines.

Dynamometers.

Electrical Engineering, Telegraphy, Electric Lighting and its practical details, Telephones

Engines, Varieties of.

Explosives. Fans.

Founding, Moulding and the practical work of the Foundry.

Gas, Manufacture of.

Hammers, Steam and other Power.

Heat. Horse Power.

Hydraulics.

Hydro-geology.

Indicators. Iron.

Lifts, Hoists, and Elevators.

Lighthouses, Buoys, and Beacons.

Machine Tools.

Materials of Construction.

Meters.

Ores, Machinery and Processes employed to Dress.

Piers.

Pile Driving.

Pneumatic Transmission.

Pumps.

Pyrometers.

Road Locomotives.

Rock Drills.

Rolling Stock.

Sanitary Engineering.

Shafting.

Steel.

Steam Navvy.

Stone Machinery.

Tramways.

Well Sinking.

London: E. & F. N. SPON, 125, Strand.

New York: 35, Murray Street.

Crown 8vo, cloth, 485 pages, with illustrations, 5*s*.

WORKSHOP RECEIPTS,
SECOND SERIES.

By ROBERT HALDANE.

SYNOPSIS OF CONTENTS.

Acidimetry and Alkali-
   metry.
Albumen.
Alcohol.
Alkaloids.
Baking-powders.
Bitters.
Bleaching.
Boiler Incrustations.
Cements and Lutes.
Cleansing.
Confectionery.
Copying.

Disinfectants.
Dyeing, Staining, and
   Colouring.
Essences.
Extracts.
Fireproofing.
Gelatine, Glue, and Size.
Glycerine.
Gut.
Hydrogen peroxide.
Ink.
Iodine.
Iodoform.

Isinglass.
Ivory substitutes.
Leather.
Luminous bodies.
Magnesia.
Matches.
Paper.
Parchment.
Perchloric acid.
Potassium oxalate.
Preserving.

Pigments, Paint, and Painting : embracing the preparation of *Pigments*, including alumina lakes, blacks (animal, bone, Frankfort, ivory, lamp, sight, soot), blues (antimony, Antwerp, cobalt, cœruleum, Egyptian, manganate, Paris, Péligot, Prussian, smalt, ultramarine), browns (bistre, hinau, sepia, sienna, umber, Vandyke), greens (baryta, Brighton, Brunswick, chrome, cobalt, Douglas, emerald, manganese, mitis, mountain, Prussian, sap, Scheele's, Schweinfurth, titanium, verdigris, zinc), reds (Brazilwood lake, carminated lake, carmine, Cassius purple, cobalt pink, cochineal lake, colcothar, Indian red, madder lake, red chalk, red lead, vermilion), whites (alum, baryta, Chinese, lead sulphate, white lead—by American, Dutch, French, German, Kremnitz, and Pattinson processes, precautions in making, and composition of commercial samples—whiting, Wilkinson's white, zinc white), yellows (chrome, gamboge, Naples, orpiment, realgar, yellow lakes) ; *Paint* (vehicles, testing oils, driers, grinding, storing, applying, priming, drying, filling, coats, brushes, surface, water-colours, removing smell, discoloration ; miscellaneous paints—cement paint for carton-pierre, copper paint, gold paint, iron paint, lime paints, silicated paints, steatite paint, transparent paints, tungsten paints, window paint, zinc paints) ; *Painting* (general instructions, proportions of ingredients, measuring paint work ; carriage painting—priming paint, best putty, finishing colour, cause of cracking, mixing the paints, oils, driers, and colours, varnishing, importance of washing vehicles, re-varnishing, how to dry paint ; woodwork painting).

London: E. & F. N. SPON, 125, Strand.
New York: 35, Murray Street.

In demy 8vo, cloth, 600 pages, and 1420 Illustrations, 6s.

SPONS'

MECHANIC'S OWN BOOK;

A MANUAL FOR HANDICRAFTSMEN AND AMATEURS.

CONTENTS.

Mechanical Drawing—Casting and Founding in Iron, Brass, Bronze, and other Alloys—Forging and Finishing Iron—Sheetmetal Working —Soldering, Brazing, and Burning—Carpentry and Joinery, embracing descriptions of some 400 Woods, over 200 Illustrations of Tools and their uses, Explanations (with Diagrams) of 116 joints and hinges, and Details of Construction of Workshop appliances, rough furniture, Garden and Yard Erections, and House Building—Cabinet-Making and Veneering—Carving and Fretcutting—Upholstery—Painting, Graining, and Marbling—Staining Furniture, Woods, Floors, and Fittings—Gilding, dead and bright, on various grounds—Polishing Marble, Metals, and Wood—Varnishing—Mechanical movements, illustrating contrivances for transmitting motion—Turning in Wood and Metals—Masonry, embracing Stonework, Brickwork, Terracotta, and Concrete—Roofing with Thatch, Tiles, Slates, Felt, Zinc, &c.— Glazing with and without putty, and lead glazing—Plastering and Whitewashing—Paper-hanging—Gas-fitting—Bell-hanging, ordinary and electric Systems—Lighting—Warming—Ventilating—Roads, Pavements, and Bridges—Hedges, Ditches, and Drains—Water Supply and Sanitation—Hints on House Construction suited to new countries.

London: E. & F. N. SPON, 125, Strand.
[New York: 35, Murray Street.